Springer Tracts in Modern Physics
Volume 128

Editor: G. Höhler
Associate Editor: E. A. Niekisch

Springer Tracts in Modern Physics

Volumes 90–111 are listed on the back inside cover

* denotes a volume which contains a Classified Index starting from Volume 36

Dieter Schumacher

Surface Scattering Experiments with Conduction Electrons

With 55 Figures

Springer-Verlag Berlin Heidelberg GmbH

Dr. Dieter Schumacher

Lehrstuhl für Oberflächenwissenschaft, Heinrich-Heine-Universität Düsseldorf
Universitätsstrasse 1, W-4000 Düsseldorf 1, Fed. Rep. of Germany

---

Manuscripts for publication should be addressed to:

Gerhard Höhler

Institut für Theoretische Teilchenphysik der Universität Karlsruhe, Postfach 69 80,
W-7500 Karlsruhe 1, Fed. Rep. of Germany

Proofs and all correspondence concerning papers in the process of publication should be adressed to:

Ernst A. Niekisch

Haubourdinstrasse 6, W-5170 Jülich 1, Fed. Rep. of Germany

ISBN 978-3-662-14947-8   ISBN 978-3-540-47474-6 (eBook)
DOI 10.1007/978-3-540-47474-6

Typesetting: Camera ready copy by author
Production Editor: P. Treiber

56/3140-543210 – Printed on acid-free paper

Preface

In surface physics it is a common procedure to scatter a beam of monoenergetic particles (electrons, atoms, ions, photons, etc.) at a solid-vacuum boundary to obtain information about surface conditions or surface processes. The aim of this book is to discover possible ways of using the surface scattering of conduction electrons of a thin metal film to investigate surface processes in a comparable way. A thin film is in this sense a sample which has in one dimension a size comparable to the mean free path of the conduction electrons. Under suitable conditions, the resistivity of such a thin film depends very strongly on its surface conditions. Even less than a monolayer coverage of an adsorbed gas or condensed metal on the smooth film surface can cause a significant increase in resistivity. For example, at a substrate temperature of 20K, the condensation of one thousandth of a monolayer of silver onto a carefully prepared thin silver film can cause a resistivity increase of approximately 0.1%, which can easily be measured with modern current and voltage meters. The high sensitivity of this method leads to a large number of interesting applications. It becomes possible to study surface processes like adsorption, desorption, surface diffusion, ordering phenomena and the first stages of crystalline growth.

Düsseldorf, May 1992   *Dieter Schumacher*

Acknowledgements

This work was prepared and compiled during my activity at the Institut für Angewandte Physik and the Lehrstuhl für Oberflächenwissenschaft of the Heinrich-Heine-Universität Düsseldorf. I thank the heads of the institutes Prof. Dr. J. Kranz and Prof. Dr. A. Otto for giving me the opportunity to work continuously on my subject. I am very grateful to Prof. Dr. D. Stark and Prof. Dr. A. Otto for many fruitful, critical and stimulating discussions. This work could only be achieved with the help of many coworkers. I thank B. Arnheiter, M. Brückner, H. Grabhorn, H. Heil, C. Holzapfel, A. Karaus, D. Körwer, J. Polomski-Keip, I. Redmann, S. Rögels, Dr. W. Schlemminger, F. Stubenrauch and A. Tonscheidt for their contributions and all my colleagues for the pleasant collaboration during the last few years. The scanning tunnelling microscope, the electron transmission microscope, and X-ray diffraction studies have been essential for this work. Generously K. Besocke (IGV, Forschungszentrum Jülich), Prof. Dr. K.V. Kowallik (Botanisches Institut, Heinrich-Heine-Universität Düsseldorf) and Dr. H. Wunderlich (Institut für Strukturchemie, Heinrich-Heine-Universität Düsseldorf) put the appropriate equipment at my disposal. I am indebted to Prof. Dr. E. Gerlach (1. Physikalisches Institut, RWTH-Aachen) and Dr. B.N.J. Persson (Institut für Festkörperforschung, Forschungszentrum Jülich) for helping me to improve my theoretical conceptions. I thank Mrs. B. Derks for executing the drawings and A. Tonscheidt for carefully reading the manuscript. Finally I should like to thank my wife Ursula and my children Barbara and Stefan, without whose indulgence and patience this work would never have been completed.

Contents

1. Introduction

For normal metals the mean free path of the conduction electrons as known from the Drude-Sommerfeld model is of the order of several tens of nm at room temperature. Under suitable conditions it is possible to prepare a thin continuous metal film with the same thickness and a well defined surface structure. The dc-resistivity of such a film can significantly be influenced by surface defects as well as adsorbates. Two main approaches to this subject are possible: First, one can try to study different surface processes like adsorption, surface diffusion, etc. by measuring the resistivity. This is very attractive especially for application, since the resistivity is an easily measurable quantity. Second, one can focus on the surface scattering process of the conduction electrons itself. This process is not well understood and a quantitative description was not available until now. However, the resistivity of a thin metal film can be influenced by a lot of phenomena. Therefore, Chap. 2 is a short review of these phenomena. It is written with the intention of giving definitions of the relevant quantities, an introduction into these phenomena, and references to extended literature. A suitable theoretical description of the correlation between the surface condition and the resistivity of a thin metal film is not available. In Chap. 3, ideas and theoretical concepts to describe the surface ·scattering process of conduction electrons and its influence on the film resistivity are collected. When studying surface physics, conductivity measurements are worthless if the metal films are not prepared under sufficiently clean conditions (e.g. evaporation under ultra high vacuum conditions). Since the metal films are at the same time "substrates" and "sensors", it is necessary to characterize them carefully by other surface- and microprobe analytical methods. These results are compiled in Chap. 4. The examples given in Chap. 5 prove that in many cases scattering experiments with conduction electrons can be used as an additional tool of surface physics.

2. The Electrical Conductivity of Thin Metal Films

The electric and electronic properties of thin metal films can differ from those of the corresponding bulk material due to a large number of phenomena. This chapter gives a review of the possible influences on the resistivity of thin continuous metal films. The supplement "continuous" means in this context, that the naive preliminary picture of a thin film which is given in Fig. 2.1 is a rough, but not incorrect, model. Generally, films with an island-, channel- or hole like structure are excluded as well as films whose electronic structure is significantly influenced by a sufficiently high defect density. The space- and velocity coordinates $\boldsymbol{r} = (x, y, z), \boldsymbol{v} = (v_x, v_y, v_z)$ and the angles φ and ϑ are also defined in Fig. 2.1. It illustrates the geometry of the measurement and

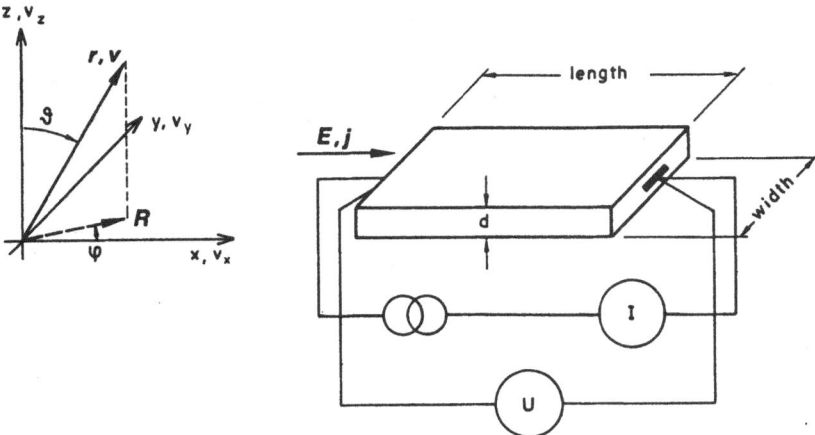

Fig. 2.1. Rough model of a thin continuous film. Definition of the space- and velocity coordinate system and the electrical current- and field directions

gives the direction of the current density $\boldsymbol{j}$ and the electric field $\boldsymbol{E}$, as well. In general, the case without external magnetic field ($\boldsymbol{B} = 0$) is considered. The (nearly) two-dimensionality and a special crystalline structure of a thin metal film can cause the electric properties to differ from those of the bulk. The possibility to observe quantum-effects directly in macroscopic quantities arises from the two-dimensionality of the thin film. Due to special preparation procedures, the thin metal films often exhibit a specific crystalline structure. The geometry of the film causes the surface to influence its electrical properties significantly, while the surface structure itself is determined by the production process.

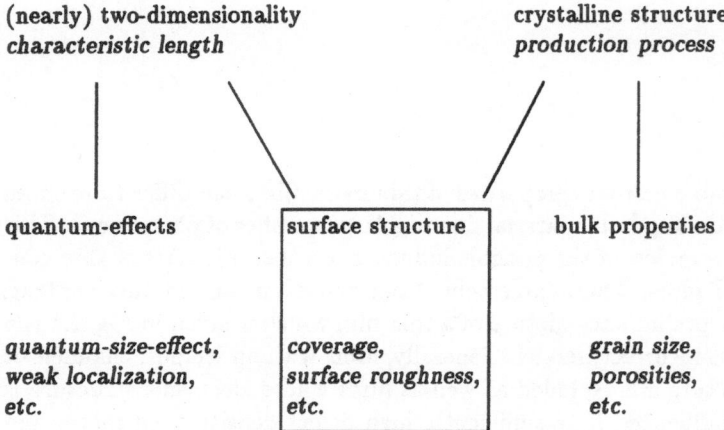

Fig. 2.2. Reasons for differences in the electrical behaviour of a thin metal film and of bulk-material

It is impossible to investigate the influence of the surface on the film resistivity isolated from these other aspects because of these relations. After a definition of the notion "two-dimensionality", short reviews of the phenomena and relationships shown in the left and right columns of Fig. 2.2 will be given.

2.1 Characteristic Lengths

It is necessary to compare the film thickness d, with the characteristic lengths (Fig. 2.3) of the conduction process. As the electric conduction is a transport phenomenon, the characteristic lengths are always connected with characteristic times For more details contact textbooks like e.g. (Ashcroft, Mermin 1981).

Within the free electron model (FEM), Ohm's law can be derived by solving the linearized Boltzmann-equation:

$$j = \sigma E, \quad \text{with} \quad \sigma = \varrho^{-1} = \frac{ne^2\tau}{m} \tag{2.1}$$

j: current density E: electric field strength
σ: conductivity ϱ: resistivity
n: electron density τ: relaxation time
m: effective electron mass e: elementary charge

The *resistivity*, $\varrho = 1/\sigma$, depends on the electron density n, the scalar effective mass m and the *relaxation time* τ. The *conductivity* σ and the effective electron mass m are generally tensorial quantities. They can often be treated as scalars, to a good approximation.

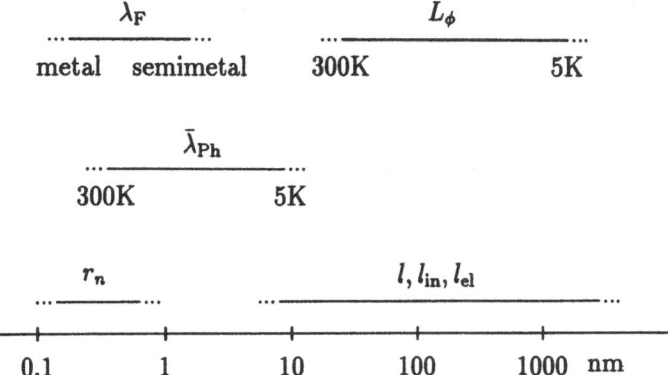

Fig. 2.3. Characteristic lengths of the conduction process in metals

l mean free path (mfp) λ_F Fermi-wavelength
l_{el} elastic mfp L_ϕ phase-coherence length
l_{in} inelastic mfp $\bar{\lambda}_{Ph}$ mean phonon wavelength
r_n interelectronic distance

The mean free path (mfp) l can be defined as the mean distance an electron travels between two scattering events. In the framework of the FEM, it can be calculated from the relaxation time τ with the knowledge of the Fermi-energy E_F or the Fermi-velocity v_F: .

$$l = v_F \tau. \tag{2.2}$$

In the FEM, the product of the mfp l and the resistivity ϱ depends on the electron density n:

$$\varrho \cdot l = \hbar e^{-2} \left(3\pi^2\right)^{1/3} n^{-2/3} \qquad \hbar = h/2\pi. \tag{2.3}$$

The majority of models concerning the conductivity of thin continuous metal films use this relation because the electron density depends only weakly on the temperature and the defect density. The electron density is concealed as a length in Fig. 2.3. The *interelectronic distance* r_n is the radius of a sphere with a volume equal to the volume per electron:

$$r_n = \left(\frac{3}{4\pi n}\right)^{1/3}. \tag{2.4}$$

One might expect deviations of the electric behaviour if the film thickness is comparable with the mfp or smaller than it. The interaction of the conduction electrons with the boundaries should influence the conduction process substantially. The concept of the mfp as a geometrical distance between two scattering events is rather vivid, but in the case of very thin or highly disordered films unsuitable.

Extending Drude's theory of conduction, Sommerfeld has replaced the classical energy distribution, originating from Boltzmann, by the Fermi-Dirac distribution function. Numerous models of film conductivity are based on Sommerfeld's approach. In most cases they differ from Sommerfeld's concept only in the boundary conditions for the thin film. He implanted a distribution function derived from a quantum mechanical concept into an otherwise classical transport theory. However, a classical description of the electron motion can only be valid, if the position r and momentum p can be stated without violating the uncertainty principle. A conduction electron with *Fermi-energy* E_F has the *Fermi-wavevector* k_F and the *Fermi-wavelength* $\lambda_F = 2\pi/k_F$. In order to use a (semi)classical approach, the uncertainty in momentum Δp must be much smaller than the electron momentum $\hbar k_F$:

$$\Delta p \ll \hbar k_F \Rightarrow \Delta r \gg k_F^{-1}. \tag{2.5}$$

In the FEM, $k_F^{-1} = (4/9\pi)^{1/3} r_n \approx 0.52 r_n$ is valid. Therefore k_F^{-1} is of the order of the interelectronic and interatomic distances (see Fig. 2.3). This means that any (semi)classical description loses its justification if it becomes necessary to state the position of an electron within a few atomic distances. In the case of an extended metal solid, Sommerfeld's procedure is justified because, generally, the mfps are larger than 10nm. However, during evaluation and interpretation of thin film data, this restriction should be kept in mind.

The mfp and the relaxation time were introduced into the FEM without an exact knowledge of the different types of contributions of scattering mechanisms to the resistivity. *Matthiessen's rule* maintains order among the different influences. The generalized form of Matthiessen's rule claims that treating the different types of scattering mechanisms as independent is a good approximation. One can then calculate the inverse relaxation time or mfp by summing up over the individual proceses (index i):

$$\frac{1}{\tau} = \sum_i \frac{1}{\tau_i} \qquad \frac{1}{l} = \sum_i \frac{1}{l_i}. \tag{2.6}$$

The assumption that the common influence of the different types of scattering processes can be calculated as a 'series connection' is an explicit or implicit precondition of many theories of thin film conductivity.

The *elastic life time* τ_{el} is defined as the mean time between two scattering events, whereby the electron momentum changes only its direction. In general the elastic scattering events are considered as phase preserving processes. In an extended solid, elastic scattering processes are mainly caused by defects and impurities. Due to a widely varying defect density or composition, the elastic mfp $l_{el} := \tau_{el} v_F$ can vary over some orders of magnitude (see Fig. 2.3).

Depending on the dimensionality a of the sample, the *diffusion constant* D_α can be deduced from the elastic life time:

$$D_\alpha := \frac{1}{\alpha} v_F^2 \tau_{el}, \qquad \alpha = 2, 3. \tag{2.7}$$

The *inelastic life time* τ_{in} is the mean time between two scattering events, whereby the electron changes its energy eigenstate. This can be caused by the emission or absorption of phonons, or by electron-electron scattering. Except at very low temperatures the first process dominates. The corresponding mfp is $l_{in} := \tau_{in} v_F$.

The *electron-phonon scattering* is generally responsible for the part of the resistivity depending on temperature. A quantitative description of the contribution of the electron-phonon scattering is given by the Bloch-Grüneisen relation. This relation is deduced from the framework of the FEM and uses the phonon density of states of the Debye-model. Whereas the resulting linear approximation in the case of higher temperatures holds generally, sometimes the T^5-dependence is not fulfilled due to umklapp-processes, a complex Fermi-surface, or the domination or influences of other scattering processes.

The *phonon wavelength* $\bar{\lambda}_{Ph}$ of the centre of the phonon spectrum can be estimated with the knowledge of the sound velocity c_s. In the case of silver ($c_s \approx 2700$m/s), the mean phonon wavelength is approximately 13nm at 10K. The phonon density of states in a thin film might differ from the bulk data which influences the electron-phonon scattering.

The *electron-electron scattering* is less important in normal metals, due to Pauli's exclusion principle, which reduces the set of allowed scattering vectors. The electron-electron scattering can have a significant influence on the conductivity in highly disordered samples or at very low temperatures.

In a metal charges induced by impurities, defects or the boundaries are screened. In order to describe the static screening of fixed additional charges it is useful to define the *screening length* k_0^{-1}. The Thomas-Fermi-model of screening shows that, for instance, the Coulomb-potential of a point charge is exponentially damped with a characteristic length k_0^{-1}. The screening length k_0^{-1} can be estimated within the free electron model:

$$k_0^{-1} = \left(\frac{3\pi^2}{16}\right)^{1/3} \left(\frac{a_0}{r_n}\right)^{1/2} k_F^{-1} \tag{2.8}$$

with

$$a_0 = \frac{\hbar^2}{me^2} \approx 0.053\text{nm}, \quad \text{Bohr's radius.}$$

Since the value r_n/a_0 lies between 2 and 6 for all metals, k_0^{-1} is of the same order of magnitude as k_F^{-1}. All free electrons participate in the phenomenon of screening by changing their trajectories. A perfect screening is not possible, because the electrons are limited to kinetic energies below E and de Broglie-wavelengths above λ_F. This leads to the *Friedel-oscillations*, which are periodic variations of the local electron density in the surroundings of charged lattice defects and the metal surface with a period length of $\lambda_F/2$.

At a given temperature the energy of a conduction electron at Fermi-level is defined with an uncertainty of approximately $\Delta E \approx k_B T$. With the help of the energy-time uncertainty relation the maximum *phase-coherence time* τ_ϕ can be estimated:

$$\tau_\phi \approx \frac{\hbar}{k_B T}. \qquad (2.9)$$

In most cases the phase-coherence time τ_ϕ is limited by the inelastic life time τ_{in}, because almost all inelastic scattering events are also phase breaking events:

$$\tau_\phi \leq \tau_{in}. \qquad (2.10)$$

With the help of the above defined diffusion constant the *phase-coherence length* L_ϕ can be derived:

$$L_\phi = \sqrt{D_\alpha \tau_\phi} \qquad \text{for} \qquad \tau_\phi \gg \tau_{el}. \qquad (2.11)$$

Typical and new physical phenomena can always be expected if the thickness of a thin film or the period length of a superlattice is comparable to or smaller than one of the described characteristic lengths.

2.2 Quantum Effects

If at least one dimension of a metallic sample is smaller than the phase-coherence length L_ϕ, it is a mesoscopic system which exhibits the possibility of observing quantum effects by measuring a macroscopic quantity, e.g. the dc-resistivity. In metallic samples the most prominent representatives of quantum effects in mesoscopic systems are the weak localization (Abrahams et al., 1979), the Aharonov-Bohm-effect (Aharonov, Bohm, 1959), and the quantum-size-effect (Sandomirskii 1967).

Weak Localization

In thin metal samples at low temperatures the phase-coherence length L_ϕ can be more than 10^3 times larger than the elastic mfp l_{el}. The electron diffuses large distances without losing its phase-coherence by inelastic scattering events. In this context, the film can be regarded as two-dimensional, if the diffusion time from one boundary to the other is small compared with the phase-coherence time. A quantum mechanical approach shows: For an arbitrary starting point, the probability of finding the electron again at this point after the time t is enhanced. This effect is called weak localization (Abrahams et al., 1979). It causes an additional resistance which has a logarithmic temperature dependence in the two dimensional case at low temperatures (Garcia, Kao, Strongin 1979; Bergmann 1984).

The interference condition which leads to this phenomenon, can be destroyed by an external magnetic field and the additional resistance disappears. In order to determine scattering times or mfp's, the weak localization can be used as a time of flight experiment with conduction electrons. A detailed description has been given in (Bergmann 1984).

Aharonov-Bohm-Effect

Modern techniques of lithography permit the splitting of a thin film into two paths and to connect it again within a smaller distance than the phase-coherence length L_ϕ (Wepp et al. 1985). Under these conditions the conductivity becomes a nonlocal phenomenon and Kirchhoff's rules are no longer valid. At the splitting point the electron wave is split into two partial waves. If both paths are equal in length, the superposition of the partial waves is in general constructive at the connection point. This condition can be destroyed by an external electric or magnetic potential, which shifts the phase of one partial wave with respect to the other. Therefore a continuously rising electric or magnetic potential causes periodic variations of the resistance of such a circuit, as has been shown by (Wepp et al. 1985; Washburn et al. 1987). In contrast to a normal field effect transistor this is a "switch" which uses the direct influence of the electric or magnetic potential on the phase of the wave function.

Quantum-Size-Effect

If the thickness of a thin film is comparable with the Fermi-wavelength of the conduction electrons, a rough quantization of the electronic levels in k-space results. This phenomenon is called quantum-size-effect (QSE). An actual review is given by (Trivedi, Ashcroft 1988). The existence of transversal standing electron waves in the film is a necessary precondition of the QSE. Therefore the phase-coherence length must be much larger than the film thickness, and the boundaries of the film must act as 'mirror like' planes. In the framework of the FEM the Fermi-body of the thin model film consists of disks, parallel to the k_x-k_y-plane and enveloped by the Fermi-sphere (Garcia, Kao, Strongin 1972). The kinetic energy of the charge carriers consists of a quasi-continuous fraction $E_\parallel$, connected with the motion in the film plane, and a roughly quantized fraction E_z ,due to the motion perpendicular to the boundaries. With increasing film thickness, one disk after another shifts into the bulk Fermi-sphere. This means that there exists a critical thickness. Below it no delocalized electronic states are possible and above it the density of delocalized states, especially near E_F , varies cyclically with the thickness. This behaviour of the density of states influences, among others, the position of the Fermi-energy, the electron density, and the surface energy (Sandomirskii 1967; Schulte 1976; Feibelman 1983; Feibelman, Hamann 1984). This should be observable as an oscillating behaviour of, for instance, the work-function, the electrical conductivity, the electronic contribution to the heat conductivity, the optical absorption and reflectivity etc.

2.3 Film Structure

Caused by the individual production process, the thin film can exhibit a special crystalline structure which differs more or less from the corresponding bulk material. Giving all details would be beyond the scope of this work, therefore the description is restricted to monocrystalline and textured pure metal films.

A monocrystalline film consists of crystallites whose height corresponds to the film thickness and whose orientation is uniformly terminated by the substrate. Such films possess small-angle grain boundaries. These are formed at planes where crystallites, which have grown independently, touch one another. The lateral extension of the crystallites is nearly identical to or larger than the film thickness.

The crystallites of a textured film lie with a selected crystalline direction parallel to one another. This axis can be seen as an infinite axis of rotation, of the set of crystallites. Generally, the crystallites approximately have a cubic shape where the edge length corresponds to the film thickness. Low-dimensional defects can be expected for both types of films inside the grains.

Production Procedure

Many different procedures exist to produce a thin metal film. Table 2.1 gives a general survey:

In fundamental research the most common procedure to prepare a thin metal film is the evaporation process. Often the films are evaporated under unsuitable vacuum conditions. Ultra high vacuum conditions with a residual gas pressure $p_r \leq 1 \times 10^{-8}$Pa are necessary to get sufficiently pure metal films.

In order to describe the electrical properties of a thin metal film in the given context it is sensible to define the quantities σ_∞, ϱ_∞ and l_∞. These are the conductivity, the resistivity and the mfp of an extended sample with the same composition and crystalline structure as the thin film. They might be compared with the corresponding values σ_{poly}, ϱ_{poly} and l_{poly} as given by

Table 2.1. Procedures to prepare a thin film

evaporation	*chemical deposition techniques*
filament source	pyrolysis
crucible source	electro-chemical deposition
electron beam source	chemical vapour deposition (CVD)
flash-heating	metal organic CVD (MOCVD)
molecular beam epitaxy (MBE)	
ion beam assisted evaporation	
sputtering	*additional physical techniques*
ion beam sputtering	gas phase epitaxy
plasma sputtering	liquid phase epitaxy
bias-sputtering	

the literature for a polycrystalline solid (Landolt-Börnstein 1959). Such a comparison is not very elucidating because it only shows the well known fact that thin metal films often have an enhanced defect density. In more detail, it is useful to distinguish between two not necessarily additive contributions to the bulk resistivity of the thin film, the *background scattering* and the *grain boundary scattering*.

Background Scattering

The conductivity, resistivity and mfp of an extended sample which has the same stucture and composition as the interior of the grains are denoted by σ_B, ϱ_B and l_B . The dominating scattering mechanisms of such a hypothetical sample are the inelastic electron-phonon interaction and the elastic scattering at low-dimensional defects. These values are often replaced by σ_{poly}, ϱ_{poly} and l_{poly}, which can only be a rough approximation.

Grain Boundary Scattering

Mayadas and Shatzkes were the first to develop a concept to describe the influence of the grain boundary scattering on the resistivity of textured thin metal films (Mayadas, Shatzkes 1970; Mayadas, Shatzkes, Janak 1973). They derived the following approximation by solving Boltzmann's transport equation:

$$\frac{\sigma_\infty}{\sigma_B} = 1 - 3\left[\alpha/2 - \alpha^2 + \alpha^3 \cdot \ln(1 + 1/\alpha)\right] \tag{2.12}$$

with

$$\alpha = \frac{l_{\mathrm{B}}}{\overline{D}} \frac{R_{\mathrm{g}}}{1 - R_{\mathrm{g}}}.$$

The parameter α depends on the mean distance between grain boundaries $\overline{D}$, the reflection coefficient of the boundaries R_{g}, and the mfp l_{B} of the background scattering. Relation (I.12) shows that the background scattering and the grain boundary scattering need not obey Matthiessen's rule and that the grain boundary scattering might cause a dependence of

$$\varrho_{\infty} = A + B\overline{D}^{-1} \qquad A, B: \text{constant values} \tag{2.13}$$

for $\alpha \ll 1$. This is of high importance, since it is a well known experimental fact that often $\overline{D} \sim d$, which causes an implicit thickness dependence (Sambles 1983; Vries 1987).

More elaborated models of the grain boundary contribution to the thin film resistivity have been given, especially by (Pichard, Tellier, Tosser 1979; Tellier, Pichard, Tosser 1979; Pichard, Tellier, Tosser 1980; Tellier, Tosser 1982; van der Voort and Guyot 1971; Warkusz 1988).

In order to find a direct correlation between surface conditions and the film resistivity, two conclusions can be drawn: First, it is sensible to minimize the influence of bulk scattering processes by seeking the preparation conditions that give a minimum of ϱ_{∞} or maximum l_{∞}. Second, it is necessary to be aware of an influence of the bulk structure on the observed surface effects.

2.4 Classical-Size Effect

For the first time, Thomson discussed the question of whether the boundaries of a thin metal film might influence its resistivity nearly one century ago (Thomson 1901). Such an influence can be expected if the film thickness is comparable to or smaller than the mfp of the conduction electrons (classical-size-effect, CSE). In a classical image the conduction electrons can either be diffusely scattered or be specularly reflected at the film boundaries. Fuchs introduced the specularity parameter p as a phenomenological quantity to describe the fraction of conduction electrons which are specularly reflected at the boundaries of the metal film (Fuchs 1938). With these boundary conditions a solution of Boltzmann's transport equation is possible (Fuchs 1938; Sondheimer 1952) and the dependence $\varrho(d, p)$ is obtained. The calculation is based on a number of rather restrictive preconditions: a single parabolic conduction band, a homogeneous and thickness independent bulk-structure, and plane-parallel boundaries. However, Fuchs' approach is rather popular to describe measured thickness dependences of the resistivity of thin metal films, even though it is often unsuitable due to the preconditions, mentioned above. Figure 2.4 shows a schematic representation of the thickness dependence of the resistivity of thin metal films (Dayal, Finzel, Wißmann 1987). This curve reveals the problematic nature of measuring and interpreting a $\varrho(d)$-characteristic. Several not clearly defined regions can be distinguished.

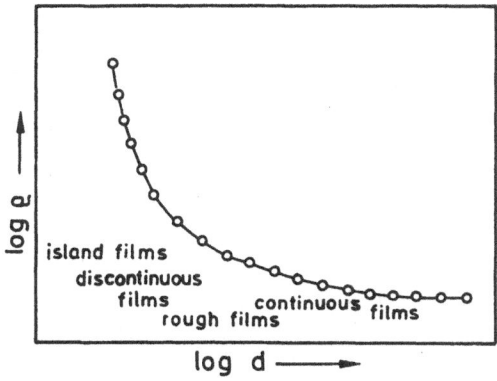

Fig. 2.4. Schematic representation of the consistently observed increase in film resistivity with film thickness (Dayal, Finzel, Wißmann 1987)

Very thin films nearly always exhibit an island-like structure. Their conductivity is dominated by the tunnelling process from island to island. In the adjacent region the films are also discontinuous, but they possess a hole- and channel-like structure and often exhibit a high surface roughness. With rising thickness the films become continuous and in a limited region a dependence like

$$\varrho = A + Bd^{-1} \qquad A, B: \text{constant values} \qquad (2.14)$$

can be observed, in agreement with simple models (Fuchs 1938; Sondheimer 1952; Mayadas, Shatzkes 1970; Wißmann 1975). In general, it cannot be decided whether this dependence is caused by scattering events of electrons at the inner grain boundaries or at the film surfaces. Also recrystallization can occur in thin continuous films due to a rising film thickness or a heat treatment, causing additional complications.

Though plenty of experimental data has been accumulated during the last decades, only a small amount of data is suitable to get a better understanding of the classical-size-effect. Only a small number of authors has been able to show that the observed thickness dependence of the resistivity $\varrho(d)$ is exclusively caused by a surface influence (Sambles 1983).

Another and in principle more suitable experimental approach to investigate the classical-size-effect was invented by (Lucas 1964). He evaporated a gold film at a residual gas pressure of 4×10^{-4}Pa and annealed it in the air at a temperature of 620K. After remounting it into the vacuum recipient, he evaporated additional gold onto the film and measured the variation of the film resistance as a function of the overlayer thickness. The resistance increase

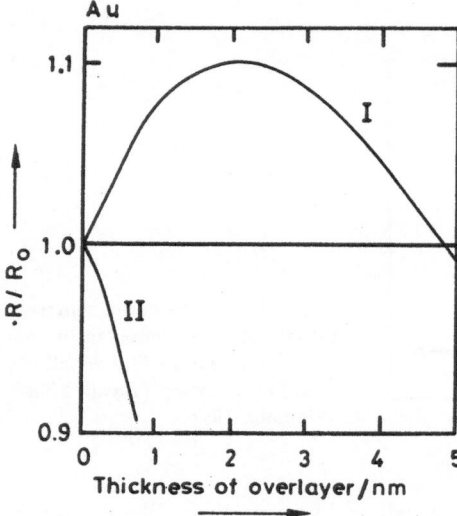

Fig. 2.5. Normalized change of the resistance R/R_0 of gold films during coverage with additional gold at room temperature. Curve I: film previously annealed at 620K, curve II: film kept at room temperature (Lucas 1964)

observed by Lucas is shown in Fig. 2.5 (curve I). Above an overlayer thickness of about 5nm, the original resistance is reobtained. If the heat-treatment is omitted a direct resistance decrease is obtained. He interprets these results in the framework of Fuchs' model; the annealed film possesses a rather smooth surface where the conduction electrons are mostly specularly reflected. The deposition of additional gold at room temperature causes a roughening of the surface at atomic scale, which leads to diffuse scattering of conduction electrons. In the unannealed film the overcoating does not change the surface structure and does not influence the surface contribution to the film resistance. The insufficient vacuum conditions and an obvious error in the base film thickness measurement cause doubts about these results (Chopra 1969). Numerous experiments of this type were done by (Chopra, Randlett 1967) and also by (Lucas 1968) in the following years with heterogeneous overcoated films. In all cases the vacuum conditions are not sufficient to exclude a contamination of the base film by residual gas molecules during the evaporation of the base film and the coating film. Chauvineau and Pariset picked up Lucas' idea and corroborated his results in principle with a large number of similar experiments which were prepared in situ and under ultra high vacuum conditions (Chauvineau, Pariset 1973; Pariset, Chauvineau 1975; Pariset, Gasgnier, Galtier 1975; Pariset, Chauvineau 1976; Pariset 1976; Pariset, Chauvineau

1978; Chauvineau 1980). In the last two decades Wißmann and coworkers (Wißmann 1975; Dayal, Finzelmann, Wißmann 1987 and references therein), as well as Wedler and coworkers (Wedler 1987 and references therein), have investigated the influence of adsorbed gases on the electrical resistance of thin metal films.

3. Concepts
to Describe the Surface Influence

The aim of this work is to observe and study surface processes such as adsorption, desorption, surface diffusion, surface reactions, nucleation etc. by measuring the dc-resistance of a thin metal film, which acts as the substrate. This aim determines the following procedure of searching for a useful approach to describe the surface contribution to the dc-resistivity. It is not intended to elaborate an extended model, which incorporates as many structural and electronic features of the thin metal film as possible, because models with this goal are often accompanied by an unwelcome number of parameters. Ideas and concepts to understand and to describe the surface contribution to the dc-resistivity shall be discussed, whereas the restriction to textured or monocrystalline films will be maintained.

3.1 The Perfect Metal-Vacuum Boundary

It is sensible to start with some considerations of the undistorted metal-vacuum boundary. You get a perfect metal surface by cutting a metallic, defect free single crystal along a plane which is parallel to and half the distance between two closed packed atomic planes. Thereby relaxation and reconstruction phenomena should be neglible.

In order to understand the interaction of a conduction electron with a defect-free metal-vacuum boundary it is necessary to pay attention to the electronic structure of the surface. The starting point is the jellium model and self-consistent density functional calculations that are used to obtain the local electron density near the surface [see for example (Lang 1973)]. Figure 3.1 is taken from a review of Lang (Lang 1973) and shows the local electron density near the jellium edge for two different interelectronic distances r_n. The z-coordinate is orientated perpendicularly to the metal surface and is normalized to the Fermi-wavelength λ_F. In front of the first ion row there is a tail of the local electron density which is evidently filled by those electrons with the highest kinetic energy, having a value of approximately the Fermi-energy (Persson, Lang 1982). The local periodic variations of the electron density in the neighbourhood of the jellium-edge are Friedel-oscillations (see 2.1). The lateral distribution of the electron density near the surface cannot be obtained from the jellium model. However, it can be determined by the

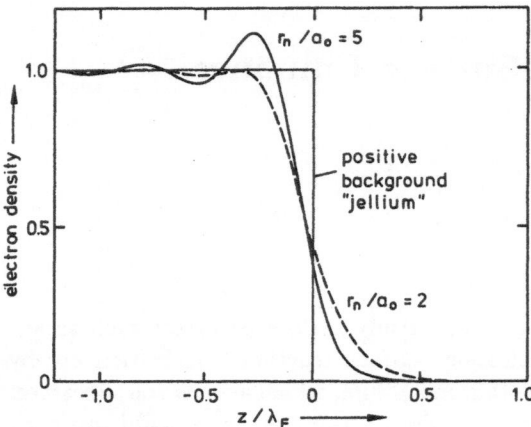

Fig. 3.1. Local electron density near the jellium-edge, obtained from self-consistent density functional calculations (Lang 1973)

self-consistent local Gaussian orbitals method as done by (Euceda, Bylander, Kleinman 1983) as well as (Arlinghaus, Gay, Smith 1980; Arlinghaus, Gay, Smith 1981). Figure 3.2 shows the contours of equal electron density at a

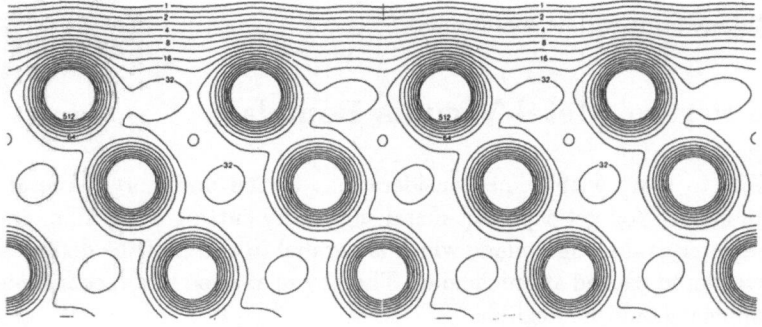

Fig. 3.2. Contours of equal electron density at a Cu(111)-surface in the (110)-plane (Euceda, Bylander, Kleinman 1983)

Cu(111)-surface. The lines differ by a factor $\sqrt{2}$. The ripple of the local electron density diminishes rapidly leaving the surface. At about one interplanar spacing in front of the uppermost ion row, the local electron density is nearly no longer corrugated.

The scanning tunnelling microscope provides the possibility to visualize the local electron density in front of a metal or a semiconductor surface. Considering the suitable polarity, the tunnelling tip follows a plane of constant local electron density to a good approximation (Garcia, Flores 1984). Governed by the tunnelling process from an occupied energy level at the surface to an unoccupied level in the tip, the microscope probes the electron density at the Fermi-level. Figure 3.3 shows a scanning tunnelling micrograph of the (111)-surface of an evaporated silver film. (For preparation see 4.1).

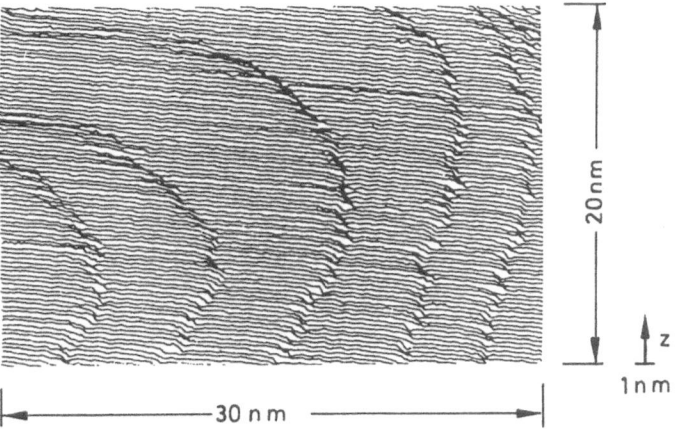

Fig. 3.3. Scanning tunnelling micrograph of the (111)-surface of a silver film, area ≈30nm×20nm

The displayed plane, lying approximately one interplanar spacing in front of the uppermost ion row, consists of extended smooth terraces separated by monoatomic steps.

The local electron density in front of the surface is connected with a surface-potential via self-consistent calculations which include exchange and correlation effects. In a single electron picture this potential forces the electrons to remain in the metal. Though exact calculations are missing, it is evident that this electron potential has a mirror-like influence on an electron striking the surface. This is a subsequent justification to describe the unperturbed surface with periodic boundary conditions in a one electron picture. The evidence proving that conduction electrons that strike a nearly perfect metal surface are indeed specularly reflected was observed among others by Koch et al. (Nee, Koch, Prange 1968; Koch, Jensen 1969; Doezema, Koch 1972) by the existance of the magnetic surface state resonances. Prange and Nee have shown that the appearance of sharp resonance lines can only be explained by repeated specular reflections (Prange, Nee 1968).

Figure 3.4 shows the resistivity of a series of thin monocrystalline silver films (Karaus 1984). The films were epitaxially grown on a Si(111) substrate. The quality of these films was checked by low energy electron diffraction (LEED) (Karaus 1984) and by transmission electron microscopy (TEM) (Polomski-Keip 1986). The solid line represents a resistivity of $1.78\mu\Omega$cm. Under the assumption that the product of the mfp and the resistivity is a constant value $[(\varrho \cdot l)_{Ag} = 843\Omega$nm^2, see above], the corresponding mfp of the conduction electrons approximately equals 47nm. Though the mfp is partly larger than the film thickness, there exists no significant thickness dependence

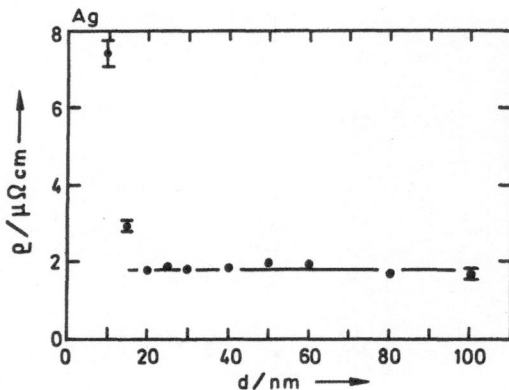

Fig. 3.4. Resistivity ϱ of thin monocrystalline silver films versus film thickness d. The mfp of the conduction electrons $\approx$ 47nm Temperature: T = 290K (Karaus 1984)

of the resistivity ϱ down to a thickness of 20nm. The large increase observed for low film thicknesses is caused by a hole- and channel-like structure of the films. This behaviour can only be explained if periodic boundary conditions are assumed.

3.2 Defects and Adsorbates

In order to find an adequate description for the interaction of the conduction electrons with surface defects and adsorbates, it is sensible to compare it to the scattering of other particles at a metal-vacuum boundary. An initial qualitative approach is derived from a comparison to the scattering of thermal helium atoms at metal surfaces (Comsa, Poelsema 1985). A loan from the light scattering was taken by Ziman (Ziman 1960). A quantitative approach can only be expected if an adequate model potential is assumed and the methods of (dynamical) LEED-calculations are applied [see for example (Pendry 1974; van Hove, Tong 1979)].

Total Scattering Cross-section of Diffuse Scattering

The flattened local electron density in front of a close-packed metal surface was the reason why the scattering of thermal helium atoms (TEAS) seemed not to be a useful tool of surface physics for a long time. In TEAS a monoenergetic beam of atoms (often helium) with a kinetic energy in the region of 10meV to 100meV, corresponding to wavelengths of about 0.14nm and 0.05nm, is scattered at a solid-vacuum boundary under ultra high vacuum conditions [for more details see (Comsa, Poelsema 1985)]. The classical turning point of the helium atom trajectories lies about 0.3nm to 0.4nm in front of the first ion row. Here the electron density shows nearly no corrugation.

Therefore for several years only the specular beam was obtained. The first order scattering signal at a Ag(111)-plane which has 10^{-4} to 10^{-3} times the intensity of the 00-reflex was observed for the first time by two groups in 1976 (Horn, Miller 1977; Boato, Cantini, Tatarek 1976). TEAS has gained in significance when it was noted that the intensity of the specular beam decreases strongly if minute amounts of adsorbates or surface defects are present. Figure 3.5 gives an example. The helium atoms are diffusely scattered at these

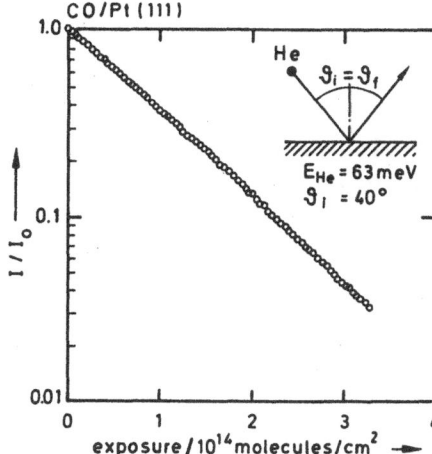

Fig. 3.5. Decrease of the relative specular intensity I/I_0 obtained from a Pt(111) surface versus CO-coverage (Comsa, Poelsema 1985)

distortions (Comsa, Poelsema 1985; Poelsema, Comsa 1989 and references therein). In a rough picture the electrons test the electron density in front of the surface from inside in a comparable way as the helium atoms do from outside. Therefore it can be sensible to borrow some ideas from this technique. At first it should be considered that thin metal films can be prepared, which fulfill Fuchs' model in a sufficient way and exhibit specularity parameters near 1. This parameter, which has the meaning of the fraction of the electrons that are specularly reflected can also be interpreted as the fraction of the surface which specularly reflects electrons. Following the TEAS approach, it is assumed that a manipulation of the smooth surface can change the interaction of the conduction electrons near a defect or impurity from a specular reflection to a diffuse scattering process. At low coverages $\theta \approx 0$, the decrease of the vacuum side specularity parameter p should be proportional to its start-value $p_v(\theta = 0)$, the coverage θ, and to the number of adsorption sites per surface area n_{surf}. The proportionality factor can be defined as the total scattering cross-section of diffuse scattering Σ:

$$\left.\frac{dp_v}{d\theta}\right|_{\theta\to 0} = -\Sigma n_{surf} p_v(\theta = 0) \qquad \theta \to 0 \qquad (3.1)$$

Sondheimer's and Juretschke's approximations, originally derived to give a simple approach of the thickness dependence of the resistivity, can be used to calculate the change of the vacuum-side specularity parameter p_v from the measured conductivity decrease (Sondheimer 1952; Juretschke 1965):

$$\varrho(p_v) \approx \varrho_\infty + \frac{3}{8}(\varrho \cdot l)_\infty \frac{1}{d}\left(1 - \frac{p_v + p_s}{2}\right) \qquad \frac{d}{l_\infty} \geq 0.1 \qquad (3.2)$$

s: substrate-side, v: vacuum-side

The value $(\varrho \cdot l)_\infty$ is assumed to be constant and well known. Sambles has shown that for noble-metal films with a (111)-texture the application of (3.2) leads to an error of $\approx 1.5\%$, due to the real shape of the Fermi-body (Sambles 1987). In general, this error is insignificant in comparison to the uncertainty in the knowledge of d, p_s, or p_v. In this context it might be sensible to define the values σ_∞ and l_∞ as the conductivity and mfp of a hypothetical film with perfectly specularly reflecting boundaries. The concept of a total scattering cross-section of diffuse scattering is similar to Wißmann's scattering hypothesis (Wißmann 1975). However, the scattering hypothesis is used with the intention of giving a rough estimation if the films obviously do not fulfill the restrictive preconditions of Fuchs' model (Dayal, Wißmann 1977), whereas it is assumed here that the films can be described within the framework of this model.

Surface Profile Function

Isolated or agglomerated adatoms as well as terrace edges, cause local distortions of the one-electron surface potential. In the case of ordinary metals the extension of these distortions is comparable with the Fermi-wavelength. This fact leads to the idea of describing the scattering of conduction electrons at a rough surface in analogy of the scattering of light at rough but non-absorbing surfaces (white sheet of paper) (Ziman 1960). A first approach including flux conservation has come from (Soffer 1967). He has introduced a profile function $\xi(x, y)$, which describes a relief where the electron-waves are locally reflected. He does not need the full knowledge of the function $\xi(x, y)$, but assumes that the profile can be decribed by a Gaussian distribution function for the heights and the lateral auto-correlation. An extended discussion of the case of different profiles for both surfaces and disappearing correlation length can be found by Sambles and Elsom (Sambles, Elsom 1982). The case of finite correlation lengths, neglected until now, might become of interest, as the scanning tunnelling microscope can be used to obtain such information about the surface.

An actual quantum mechanical realization of the concept of a profile function has been given by (Leung 1984). In the framework of the FEM the profile function describes the position of the potential step at the vacuum-film- and at the substrate-film boundary:

$$U(x, y, z) = \begin{cases} 0 \text{ for } \xi_s(x, y) < z < \xi_v(x, y) + d \\ U_0 \text{ elsewhere} \end{cases} \qquad (3.3)$$

The solution of Schrödinger's equation is a very difficult task apart from some trivial and at the same time uninteresting cases. Leung simplifies the problem by rectifying the film with the help of a coordinate transformation, which leads to smooth and parallel boundaries at $z' = 0$ and $z' = d$. As a direct consequence of this nonconformal transformation the motion of the electron is not the motion of a free particle anymore, but a mass tensor dependent on position results. Leung used Soffer's parametrization of the profile function and obtained an additional film resistance, which is proportional to $\hbar$, hence of quantum mechanical origin.

Scattering at a Model-Potential

Fuchs' specularity parameter is less suitable to implement the results of a quantum mechanical scattering calculation. In 1966 Greene first suggested adequate boundary conditions (Greene 1966). Translated into a modern notation he has defined the function $w(\boldsymbol{k}, \boldsymbol{k}')$, which gives the probability that an electron which strikes the surface with the wave-vector $\boldsymbol{k}$ leaves it with the wave-vektor $\boldsymbol{k}'$. Thus the interaction is assumed to be elastic, $|\boldsymbol{k}| = |\boldsymbol{k}'|$. Since in this work the case of specular reflection is of special interest, the dynamic probability of specular reflection $w(\boldsymbol{k})$ is split off:

$$w(\boldsymbol{k}, \boldsymbol{k}') = w(\boldsymbol{k}) \cdot \delta(k_x - k_x') \cdot \delta(k_y - k_y') \cdot \delta(k_z - k_z') + W(\boldsymbol{k}, \boldsymbol{k}'). \qquad (3.4)$$

The first term describes the process of specular reflection, whereas the second term $W(\boldsymbol{k}, \boldsymbol{k}')$ gives the probability of diffuse, but not necessarily isotropic scattering. In a later publication, Greene and O'Donnel calculated the differential scattering probability in Born's approximation (Greene, O'Donnell 1966). They used an infinite barrier to describe the smooth and clean surface and the surface charge, caused by an adparticle, is represented by a screened Coulomb-potential. A lot of similar work was done by different authors (Watanabe 1973; Watanabe, Hiratuka 1979; Greene, Malamas 1973; More, Lessie 1973), but Born's approximation turned out to be unsuitable for the case of conduction electrons and a finite surface density of adparticles. Lessie and Crosson (Lessie 1979; Lessie, Crosson 1986) were the first who extended the Born-procedure and incorporated interference and multiple scattering effects using the average t-matrix approximation (ATA). Since this model eliminates the shortcomings of its precursors the main aspects shall be referred here:

Lessie and Crosson have followed the procedure of Fuchs and Lucas (Fuchs 1938; Lucas 1965) but have not suppressed the angular dependence of the specular parameters $p_v = p_v(\varphi, \vartheta)$ and $p_s = p_s(\varphi, \vartheta)$:

$$\sigma\big(p_v(\varphi, \vartheta)\big) = \sigma_\infty - \frac{3\sigma_\infty}{4\pi\kappa}\mathcal{F}(\kappa, p_v, p_s), \qquad \kappa := d/l_\infty \qquad (3.5)$$

with

$$\mathcal{F}(\kappa, p_{\rm v}, p_{\rm s}) \quad := \quad \int\limits_{0}^{2\pi} \int\limits_{0}^{\pi/2} \cos\varphi \sin\vartheta \cos\vartheta$$

$$\times [2 - p_{\rm v} - p_{\rm s} + (p_{\rm v} + p_{\rm s} - 2p_{\rm v}p_{\rm s})\exp(-\kappa/\cos\vartheta)]$$

$$\times [1 - \exp(-\kappa/\cos\vartheta)][1 - p_{\rm v}p_{\rm s}\exp(-2\kappa/\cos\vartheta)]^{-1}d\varphi d\vartheta .$$

The aim of Lessie's and Crosson's approach is to replace the statistic-mechanical parameters $p_{\rm v}(\varphi, \vartheta)$ and $p_{\rm s}(\varphi, \vartheta)$ by quantum mechanical quantities. They represent the uncovered surface by a potential step of height V_0 equivalent to the difference between the Fermi-level and the vacuum level. Here a coordinate system differing from Fig. 2.1 is used. The boundary (potential step) is positioned at $z = 0$ and the film is, for the present, infinitely extended in the negative z-direction. Governed by the potential step, the one-electron wave-function $\Phi_{\boldsymbol{k}}(\boldsymbol{r})$ at $E_{\rm F}$ has the form:

$$\Phi_{\boldsymbol{k}}(\boldsymbol{r}) = \begin{cases} 2\exp(i\delta) & \sin(k_z - \delta) & \exp(i\boldsymbol{k}_{\|}\boldsymbol{r}), & z \leq 0 \\ T\exp(-\gamma_z z)\exp(i\boldsymbol{k}_{\|}\boldsymbol{r}), & z \geq 0 \end{cases} \qquad (3.6)$$

with

$$\gamma_z^2 := (2mV_0/\hbar^2) - k_z^2, \qquad \tan\delta := k_z/\gamma_z .$$

There k_z and $\boldsymbol{k}_{\|}$ are the components of the wave-vector perpendicular and parallel to the surface. In order to simplify the calculation the potential step at $z = 0$ is replaced by an infinite barrier ('hard wall') a distance Δ in front of the surface. The distance Δ is always chosen in such a way that the new wave-function remains unchanged in the interesting region $z < 0$. The adparticles or defects are represented by distortion-potentials ν_i which are statistically placed on the points of a two dimensional surface lattice. Figure 3.6 visualizes this scattering geometry for a simple cubic surface lattice. As a simplification, Lessie and Crosson assumed that the distortions are s-wave scatterers. This restriction might be overcome in the future. The decisive methodical advance of this work is the application of the average t-matrix approximation. This approximation is equivalent to the exchange of the stochastic array of perturbing potentials ν_i to an ordered array of fictitous potentials η_i (see Fig. 3.6). With the help of the obtained scattering amplitudes, the authors calculate the probability of specular reflection $\hat{p}(\boldsymbol{k}, \theta)$ and diffraction $\tilde{p}(\boldsymbol{k}, \boldsymbol{G}, \theta)$ as a function of coverage θ. Both depend on the wave-vector of the incoming electron $\boldsymbol{k}$. $\boldsymbol{G}$ is an arbitrary reciprocal lattice vector of the two-dimensional surface lattice. Since the magnitude of the wave-vector is conserved $|\ \boldsymbol{k}\ | = |\ \boldsymbol{k}'\ |$, the angles φ and ϑ can be derived from $\boldsymbol{k}$. Lessie and Crosson identified the quantum mechanical probability $\hat{p}(\boldsymbol{k}, \theta)$ with the specularity parameters $p_{\rm v}(\varphi, \vartheta)$ and

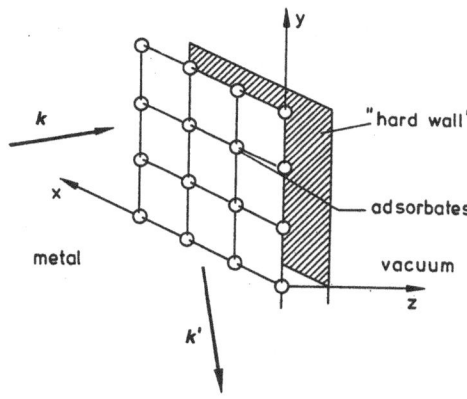

Fig. 3.6. The scattering geometry (Lessie 1979)

$p_s(\varphi, \vartheta)$. Fuchs' and Lucas' boundary conditions do not permit the consideration of a diffracted beam. The authors use an approximation: They add the probability of diffraction to the probability of specular reflection if the diffraction angle differs less than 10° from the reflection angle. The authors have done exemplary calculations for the Cu(100) surface. Some of the results are compiled in Fig. 3.7. There the substrate film boundary, as well as the vacuum film boundary is covered with scattering centres. The corresponding values are θ_s and θ_v. The resistivity ϱ_0 is the value corresponding to $\theta_v = 0$ and a certain "coverage" θ_s of the substrate-film boundary. Figure 3.7 proves that

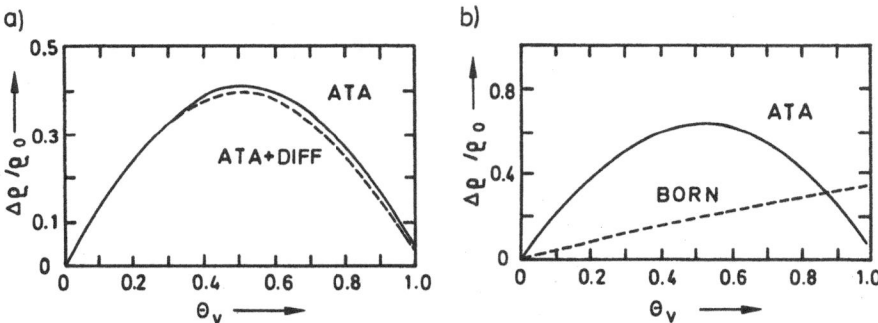

Fig. 3.7. Relative change of the resistivity $\Delta\varrho/\varrho_0$ versus coverage θ_v, $\theta_s = 0.9$. A comparison of the results: a) with the average t-matrix approximation (ATA) and in Born's approximation (BORN), $\kappa = 0.1$, b) with (ATA+DIFF) and without (ATA) consideration of the diffracted beam $\kappa = 0.3$ (Lessie, Crosson 1986)

the consideration of multiple scattering and interference effects is substantial. Born's approximation leads to roughly diverging results. However, the diffraction of conduction electrons at the lattice of distortion potentials might be negligible. An analysis of the resulting formula shows that the authors obtain a parabolic dependence in a very good approximation

$$\Delta\varrho/\varrho_0 \sim \theta_v(1 - \theta_v) \tag{3.7}$$

which is well known from the behaviour of binary alloys and is called "Nordheim behaviour".

Energy Loss Method

Procedures to calculate the resistivity are generally based on Ohm's law. A specific scattering time or length is determined in order to obtain the correlation between electric field and current density, or in other words, the resistivity. In contrast Gerlach's concept (Gerlach 1984) is based on the equivalence of energy loss and Joule heat (Gerlach, Grosse 1977). As limiting cases he studied the effect of an ion near the surface of a conducting half-space and of an ion inside and outside a very thin film. The ion is considered to be moving with a constant velocity parallel to the boundaries. In order to describe the screening of the ion charge, which causes the interaction energy between the ion and the metal, he adapted the method of Heinrichs (Heinrichs 1973). For that purpose the dielectric function $\varepsilon(k, \omega)$ must be known. Recently, Gerlach has expanded his calculations to adsorbed dipoles (Gerlach 1990).

A very promising approach to calculate the resistivity increase due to different adspecies was recently given by Persson (Persson, Schumacher, Otto 1991; Persson 1991). He related the resistivity increases to the life time $\tau_{e\text{-}h}$ of the adsorbate vibration parallel to the film surface (the parallel frustrated translation). Whereby he assumes that its damping is mainly due to the excitation of electron hole pairs. For this purpose he treats the vibration as an oscillation

$$x = x_0 \cos \Omega t \quad \text{with} \quad x_0 = 2\sqrt{\frac{\hbar}{2M\Omega}} \tag{3.8}$$

with the frequency Ω, the amplitude x_0 and the mass of the adparticle M. The energy transfer per unit time from the adsorbate to the metal film due to e-h pair damping is

$$P_\downarrow = n_{\text{surf}} A\hbar\Omega\tau_{e\text{-}h}^{-1} \tag{3.9}$$

where n_{surf} is the surface density of the adparticles and A is the covered surface area. Persson changes the reference frame to a frame which is oscillating with the same amplitude and frequency as the adsorbate. In this frame the adsorbates are stationary, whereas the electron sea oscillates with the amplitude x_0 and the frequency Ω. The electron fluctuations are connected with a current j in x-direction (e: elementary charge)

$$j = nev, \tag{3.10}$$

where n is the density of the conduction electrons and v their drift velocity. The additional surface scattering processes of the conduction electrons give rise to additional Ohmic heating. This energy transfer per unit time $P_\uparrow$ is given by

$$P_\uparrow = < j \cdot E > Ad = < j^2 > Ad/\sigma \tag{3.11}$$

$$= (nex_0\Omega)^2 Ad$$

where σ is the conductivity at the frequency Ω which, however, is essentially identical to the dc conductivity due to the low frequency of the frustrated translation (typical $\hbar\Omega \approx 5$meV). The brackets $< \dots >$ stand for averaging over time. From the equivalence $P_\downarrow = P_\uparrow$ the initial slope of the resistivity increase due to adparticles can be related to the vibrational life time $\tau_{\text{e-h}}$:

$$\left.\frac{\partial \varrho}{\partial n_{\text{surf}}}\right|_{n_{\text{surf}}\to 0} = \frac{M}{n^2 e^2} \ \frac{d}{\tau_{\text{e-h}}}. \tag{3.12}$$

Now the problem is transfered to the knowledge of the quantity $\tau_{\text{e-h}}$ for a certain adparticle. However, Persson has derived the quantitative descriptions for the electron-hole pair damping of the frustrated translation for three limiting cases of adsorbate bonds (covalent bond, ionic bond and van der Waals bond). For more details see (Persson 1991).

Film Thickness Inhomogenities

The heights of the crystallites of a textured or monocrystalline thin metal film can deviate by about 10%. This can be proven by transmission electron microscope replica and scanning tunnelling images as well as X-ray diffraction fine-structure studies (see Sect.4.2). The crystallites with lower heights contribute more than proportionally to the resistance and its changes caused by surface effects. This result does not depend on the applied model. Therefore it is reasonable to calculate the film resistivity $\varrho'(d)$ by an averaging process such as

$$\varrho'(d) = \frac{\int \varrho(d - \zeta)f(\zeta)d\zeta}{\int f(\zeta)d\zeta} \qquad \zeta \ll d \tag{3.13}$$

where $f(\zeta)$ is the probability of finding a region with height $d - \zeta$ and $\varrho(d - \zeta)$ is the resistivity of this region. Such a procedure is allowed under the assumption that the lateral extension of the regions is much larger than the Fermi-wavelength l_F and the deviation ζ is much smaller than the mean thickness d. In the case of higher deviations ζ it is necessary to apply the much more sophisticated network theory as Elson and Sambles (Elsom, Sambles 1981) have done.

Nearly all models to derive the thickness dependence of continuous metal films lead to an approximately hyperbolic function:

$$\varrho(d) \approx A + Bd^{-1}. \tag{3.14}$$

On the basis of this equation and assuming a symmetric function $f(\zeta)$ the integral can be evaluated. Expanding the result in a series, only terms with negative odd powers of d are obtained. If the series is terminated after the third term it follows (Dayal, Finzel, Wißmann 1987):

$$\varrho'(d) \approx A + Bd^{-1} + Cd^{-3}. \tag{3.15}$$

The averaging leads to an additional term Cd^{-3}. The factor C depends on the special choice of the distribution function and is connected with the mean value of ζ. With decreasing thickness $d \rightarrow 0$ the resistivity always increases more steeply than d^{-1}. There the additional d^{-3} dependence might be a suitable approximation. However, the progressive increase might not only be caused by a constant thickness inhomogenity but also by holes, channels, a varying surface roughness, recrystallisation processes, etc.. Therefore it seems to be rather doubtful to determine C or corresponding values by fitting a formular equivalent of (3.13) to experimental data as proposed by Namba (Namba 1970). This is also valid for more elaborate models. However, X-ray diffraction techniques and the scanning tunnelling microscope deliver the distribution of the crystallite heights. These data can be used to estimate the error in calculating the surface influence

3.3 Superimposed Films

First, some considerations concerning the conductivity of superimposed metal films should be described. In this context the quantities such as thickness d, resistance R, resistivity ϱ and conductivity σ are marked by index numbers. The index 0 belongs to the (still) uncovered base film, whereas the index numbers 1 and 2 refer to the base film and the covering film as parts of the double layer. If the quantities are used without an index, they belong to the whole "sandwich". A formal application of Kirchhoff's rule to the double layer leads to (Fischer 1980):

$$\frac{R_0}{R} = \frac{\varrho_0}{\varrho_1} + \frac{\varrho_0}{\varrho_2}\frac{d_2}{d_1} \tag{3.16}$$

$$\frac{\sigma}{\sigma_0} = \frac{\sigma_1}{\sigma_0}\frac{d_1}{d} + \frac{\sigma_2}{\sigma_0}\frac{d_2}{d} \quad \text{with } d = d_1 + d_2. \tag{3.17}$$

The values R_0, ϱ_0 and σ_0 can be measured directly and are assumed to be known. In the given context, $R(d_2)$ might be measured in order to get information about the growth process of the covering film. This resistance variation is a superposition of the resistance variation of the base film $\varrho_1(d_2)$ due to its coverage and the resistance variation of the coating film $\varrho_2(d_2)$ due to its growth. Equation (3.16) shows that R_0/R meets a linear dependence if ϱ_1 and ϱ_2 do not depend on d_2. A general description on the basis of (3.16) or (3.17) requires the knowledge of the functions $\varrho_1(d_2)$ and $\varrho_2(d_2)$ while considering the electron exchange between both films. A more general description, which incorporates previous models (Lucas 1968; Belzak, Kedro, Pevala 1974; Dimmich, Warkusz 1983; Dimmich, Warkusz 1986) was given by (Dimmich 1988). The covering film can be discontinuous and island-like. Therefore Dimmich incorporated a concept of Mitchinson and Pringle (Mitchinson, Pringle 1971) into his model. Following them, the covering film consists of small cubic crystallites which have an actual height of d_2 and cover a fraction of the surface, while the mean covering film thickness is d_2'. Descriptions of this type are of little use because nearly all of the numerous parameters of such models cannot be checked by independent measurements nor can they be calculated from principle concepts.

4. Thin Metal Films on Glass Supports

In order to study surface processes with the help of dc-resistivity measurements at thin films, the films must be "sensors" and "substrates" at the same time. There are many requirements that result from this application. The surface should be smooth and well defined, especially at atomic scale, and it should be built up by a close-packed atomic plane. The films must be produced and kept under ultra-high vacuum conditions to warrant an adsorbate free metal-vacuum boundary. Surface reconstruction as well as strong substrate-film interaction are unwelcome since they might cause trouble when explaning the obtained results. A weak substrate-film interaction is also desirable to avoid mechanical stress in the film during extended heating and cooling circles. In order to reach a high sensitivity and to apply Matthiessen's rule a high dc-conductivity is required. If a cleaning of the film surface is not possible, the experiment must always be done with *new* films. Therefore a high reproducibility in terms of the preparation of the films is obligatory. In order to measure the film resistance it is necessary to use a dielectric substrate. Extended preparatory work is necessary to find suitable preparation conditions and to characterize the film sufficiently. The results are shortly decribed with the example of silver films in the following section.

4.1 Preparation

The film preparation and the following surface experiments were exclusively done under ultra high vacuum conditions (residual gas pressure: $p_r \leq 1 \times 10^{-8} Pa$). The films were evaporated by tungsten crucibles. The film thickness is measured by a quartz crystal thickness monitor. Technical details of the resistivity measuring equipment will be given in Sect. 4.2. By using special glass supports and carefully chosen preparation conditions it is possible to produce textured films, which widely fulfill the conditions that are demanded above.

Substrates

Commercially fabricated glass slides are used as supports. They are produced by pressing the melted glass through a platinum nozzle (DESAG 1989). Directly after passing the nozzle the glass has an extremely smooth surface due to the surface tension. The congealing glass is transported by two counter rotating graphite rollers. The glass consists of 65% SiO_2. The remaining 35% are admixtures of different metal oxides ($Na_2O, K_2O, ZnO, Al_2O_3, B_2O_3, Sb_2O_3$ and TiO_2) (DESAG 1989). An X-ray fluorescence analysis determined this composition. But an X-ray microprobe analysis shows that S, Cl and P are enriched in a surface layer of 0.1mm depth.

Using a primary energy of about 1.1keV and a beam current below 10mA, Auger-electron-spectra can be obtained from these glass slides. The spectra were slightly shifted (10eV - 20eV) due to surface charges. However, the significant shape of certain Augerlines permits a definite assignment. Therefore it becomes possible to investigate the border of the glass supports and the growth mechanism by Auger-electron-spectroscopy (AES). The AES signal shows that the glass surface is covered by several layers of carbon graphite, which obviously originates from the rollers mentioned above. If this film is removed by a sputtering procedure, one finds a layer which contains S, Cl and F, as expected. Below this layer the composition meets approximately the known bulk values. The graphitic surface layers cause a weak conductivity which permits the Auger-spectroscopy but does not disturb the dc-resistivity measurement. Since the graphitic substrate surface is quite welcome, the glass slides are only *mechanically* cleaned by rinsing them with distilled water to remove dust particles. Additionally the supports are heated for several hours at 500K under ultra high vacuum conditions. With the help of Auger-electron-spectroscopy it can be shown that the adsorbed gases N and O_2 desorb during this heat treatment.

The Growth Mechanism

Since the Auger-electron signal only comes out of the uppermost two or three atomic layers, AES can give information about the growth process of the metal films (Argile, Rhead 1989). Figure 4.1 shows the normalized peak to peak heigths $I_{pp}/I_{pp\,max}$ of the Auger-lines of silver (351eV) and carbon (271eV) versus the mean film thickness d_Q as measured by the quartz crystal thickness monitor.

The intensities $I_{pp}/I_{pp\,max}$ can be interpreted as the fractions of silver or carbon on the surface. Figure 4.1 shows that at a mean thickness of 9nm about 50% of the surface is covered with silver nucleids. If the film thickness is greater than or equal to 20nm no carbon signal can be seen which proves that those films are dense and possess no holes. This behaviour is typical of a growth process governed by a weak substrate-metal interaction. The AES studies prove that no ingredients or impurities of the glass supports diffuse into the films or to their surface. So, especially the metal-vacuum boundary is clean and available for surface experiments for several hours.

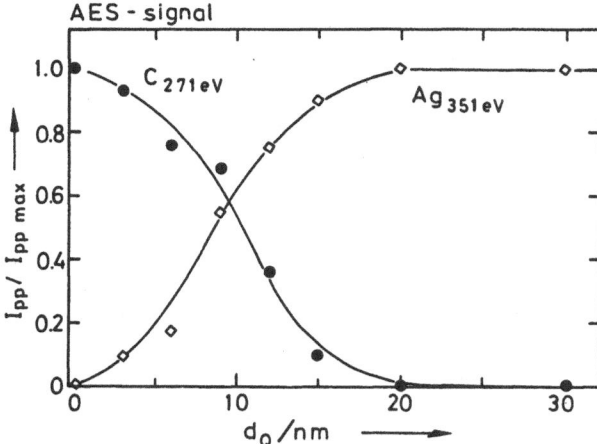

Fig. 4.1. Normalized peak to peak intensity $I_{pp}/I_{pp\,max}$ of the Auger-lines of silver (351eV) and carbon (271eV) versus the mean thickness d_Q as measured by the quartz crystal thickness monitor.

The Optimum Preparation Conditions

During the evaporation process a number of defects are incorporated into the metal film. It is sensible to distinguish between defects which can be eliminated during a following heat treatment and those which reside until the film changes substantially into an island-like structure at a certain temperature. The dc-resistivity of a metal film is under suitable conditions a sensitive indicator of the defect density in the bulk and at the boundaries. Therefore it is sensible to vary all evaporation parameters in order to find the preparation conditions which lead to a maximum dc-conductivity. They should also ensure a minimum of crystallographic defects.

Minimalization of the Residual Defects

The evaporation parameters to consider are at first the deposition rate N_D and the substrate temperature T_D during the film growth. However, the residual gas pressure p_r and its composition, the evaporation geometry, and the type of evaporation source might also influence the film properties. An annealing process is necessary to obtain films with a maximum conductivity. The films are heated with a constant rate of $r_A = 0.1 K/s$ up to the annealing temperature T_A. The thermal treatment is done in situ and is monitored by continuously measuring the film resistivity. Figure 4.2 shows the typical resistivity behaviour of an evaporated silver film. A first schematic drawing was given by Vand (Vand 1942). Structures in the irreversible part of the resistivity can give information about the energy-spectrum of the annealed

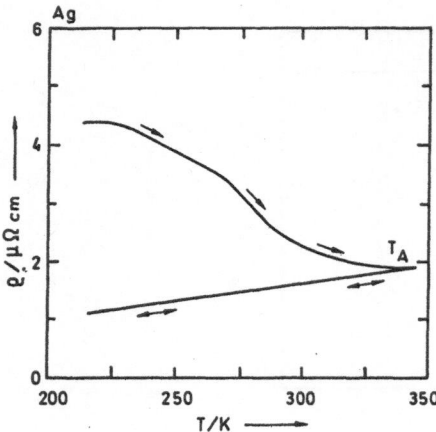

Fig. 4.2. Typical irreversible (→) and reversible (↔) temperature dependence of the resistivity of a silver film of 20nm thickness, evaporated at 225K

micro-distortions (Schumacher, Stark 1986). When the film is heated for the first time after the deposition, many of the defects incorporated during the growth are annihilated. This leads to an irreversible resistivity decrease. At a specific temperature T_A the irreversible resistivity decrease meets a minimum. Generally the heating is stopped at this point. Additional heating would not result in a lower resistivity or a more ordered structure since at a certain temperature just above T_A the film bursts and gets an island like structure. For silver films, T_A is about 340K. Below T_A the film now exhibits the well known linear and reversible temperature dependence.

The influence of the evaporation rate and the deposition temperature are not independent. But here, the conductivity obtained after the thermal treatment is a rather simple function of both parameters. This permits a separate discussion of both parameters. Figure 4.3 shows the conductivity of a series of silver films as a function of the deposition temperature T_D. The thickness equals 20nm and the evaporation rate 0.3nm/s. The resistivity directly after the evaporation σ_D, measured at 100K, increases with increasing deposition temperature due to a more ordered crystalline structure. At 350K the conductivity is lower by some orders of magnitude, because such films exhibit an island-like structure. At the same measuring temperature, the situation changes significantly after the heat treatment. Now the conductivity σ_A at 100K reaches a prominent maximum at a deposition temperature of 225K. The resulting conductivity lies just under the conductivity of the polycrystalline bulk material (see Fig. 4.3, solid line) (Landolt-Börnstein 1959). It must be emphasized that all films were equivalently heated to $T_A \approx 340$K. The temperature which leads to the maximal conductivity is called optimal deposition temperature T_D^*.

The deposition rate N_D does not influence the remaining defects significantly. This is shown in Fig. 4.4. In a wide range the conductivity obtained

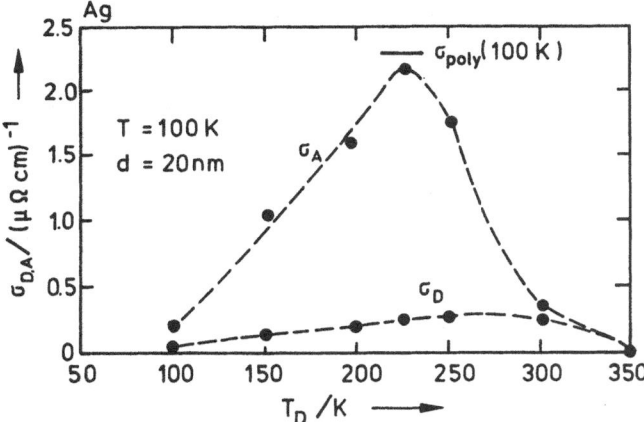

Fig. 4.3. Electrical conductivity before and after the heat treatment σ_D, σ_A. Both are measured at 100K as a function of the deposition temperature T_D for a series of silver films with a thickness of $d_Q = 20$nm

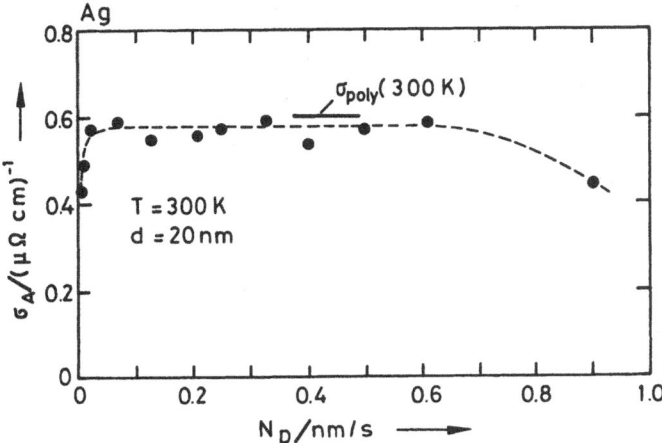

Fig. 4.4. Conductivity after the heat treatment as a function of the evaporation rate N_D of silver films with a thickness $d_Q = 20$nm at 300K

after the annealing process σ_A at 300K does not depend on the deposition rate N_D, which includes a rather constant defect density. The deviations at very high or low rates might be caused by technical problems.

A comparable optimal deposition temperature occurs for copper $T_D^* = 310K$, (Schlemminger 1989) and gold $T_D^* = 250K$. The fact that this phenomenon is not limited to fcc-metalls can be proven with the help of Fig. 4.5. In the case of gallium, which has an orthorombic crystalline structure, the optimum deposition temperature T_D^* is about 100K (Brückner 1982). For indium films a corresponding maximum of the conductivity lies at $T_D^* = 110K$.

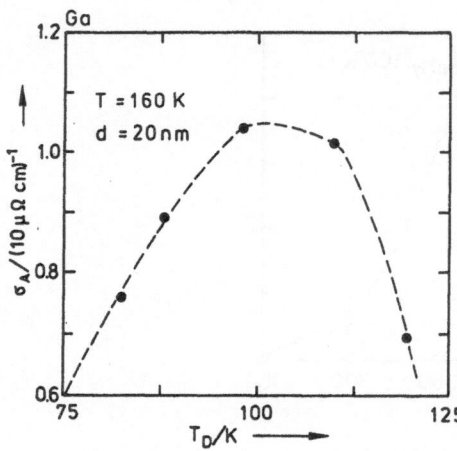

Fig. 4.5. Electrical conductivity after the heat treatment σ_A measured at 160K as a function of the deposition temperature T_D for a series of gallium films with a thickness of d_Q = 20nm (Brückner 1982)

In summary it can be stated: During the evaporation procedure a number of defects are incorporated into the thin metal film. Those types of defects which cannot be eliminated by a following heat treatment are nearly independent of the deposition rate N_D but show a significant minimum at a certain deposition temperature T_D^*. Thereby, a correlation between the temperature corresponding to the maximum of the phonon density of states and T_D^* could be observed (Schlemminger 1989). This result correlates with observations by (Bülow, Buckel 1956), but a satisfactory explanation has not been found until now.

4.2 Characterization

Under suitable evaporation conditions one obtains thin metal films on glass supports which show nearly the same conductivity as polycrystalline bulk material σ_{poly}. The following investigations show that these films are exceptionally suitable for monitoring surface processes via conductivity measurements.

4.2.1 Electron Microscopy and Electron Diffraction

Transmission electron microscopy is the standard procedure to study the morphology and crystalline structure of thin evaporated films. The investigations described in the following were done with a transmission electron microscope type EM300 (Siemens).

Sample Preparation

The commonly used glass supports are not irradiatable by the electron beam. Therefore a procedure was developed to lift the films from the glass substrate. Here the weak interaction between substrate and metal film is helpful. First an organic primer was dropped onto the film. After hardening the primer, the coating was stripped off together with the metal film and deposited on a small mesh. Now the primer is slendered in an atmosphere of a suitable organic solvent until it exhibits holes. At the borders of the holes, the thin film is now partially exposed and can be irradiated by the transmission electron microscope. The high reproducibility and the consistency with the results obtained from other methods show that this special detatching technique conserves the main features of the films. The results described in the following section concern silver films, prepared under the conditions mentioned above (T_D^*=225K, N_D=0.3nm/s, r_A=0.1K/s and T_A=340K).

Morphology

The Figs. 4.6a-c show bright-field images of three silver films with the thicknesses d_Q=15, 20, 40nm. An area of about $1.5\mu m \times 1\mu m$ is presented. The films are micro-crystalline, whereas the lateral extension of the crystallites is of the same order of magnitude as the film thickness. Especially, it can be noticed that the film with a mean thickness of $d_Q = 15nm$ possesses about 15% holes. Films with a thickness equal or above 20nm are dense which means that the fraction of holes is beneath 2%. The mean lateral extension of the crystallites $\overline{D}_{lat}$ is 1.2 to 1.7 times the mean thickness d_Q in the thickness range between 15nm and 200nm. This relation is also valid for the most frequent lateral crystallite size $\hat{D}_{lat}$.

Electron Diffraction in the Transmission Electron Microscope

The diffraction patterns of the represented micrographs (Figs. 4.6a - 4.6c) are shown as insets. In each case, roughly the whole area represented in the micrograph has contributed to the diffraction pattern. Therefore the observed ring system consists of more or less spots, due to the mean crystallite size. With counting from inside to outside the rings can be attributed to the following planes (111), (200), (220), (311), (222), (331), (420). This diffraction pattern is typical of a (111)-texture. A small lattice contraction which decreases with increasing thickness can be deduced. This result is not supported by the X-ray diffraction and can be an artifact of the stripping procedure.

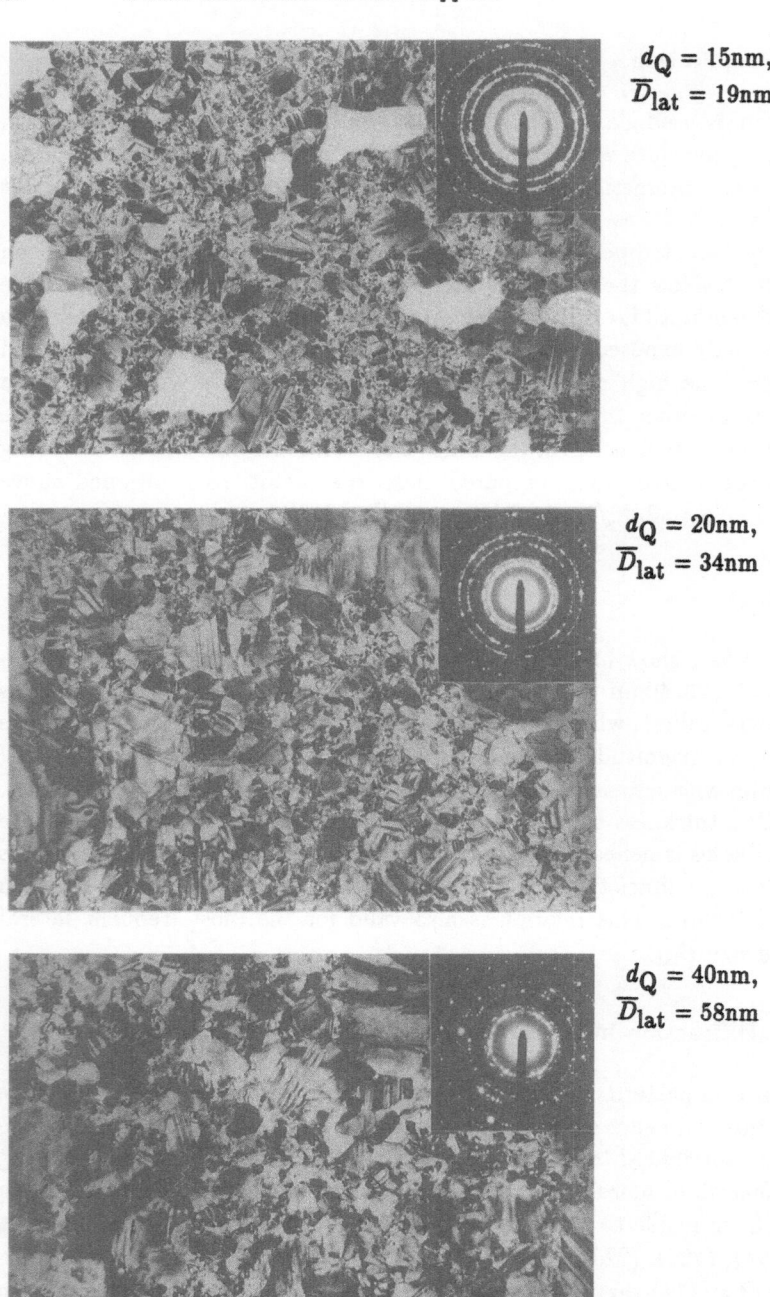

$d_Q = 15\text{nm},$
$\overline{D}_{\text{lat}} = 19\text{nm}$

$d_Q = 20\text{nm},$
$\overline{D}_{\text{lat}} = 34\text{nm}$

$d_Q = 40\text{nm},$
$\overline{D}_{\text{lat}} = 58\text{nm}$

Fig. 4.6. Transmission electron microscope images of thin silver films, area: $1.5\mu\text{m} \times 1\mu\text{m}$. Inset: diffraction patterns; mean thickness: d_Q; mean lateral crystallite size: $\overline{D}_{\text{lat}}$

Grain Boundaries Distribution

A high reproducibility of the electrical properties of the thin metal films can only be reached by carefully controlling the crystalline film structure. Especially the density of grain boundaries can influence the electrical conductivity of a thin film significantly. It must be ensured that the grain boundary dis-

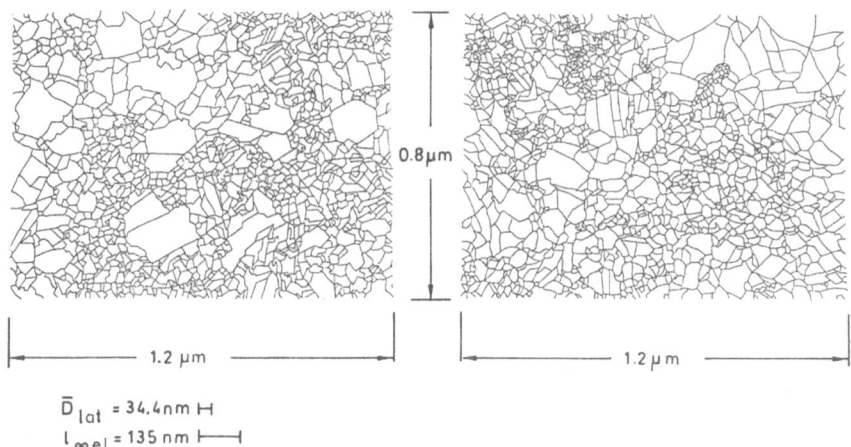

0.8 μm

1.2 μm   1.2 μm

$\overline{D}_{lat} = 34.4\,nm$ ⊢⊣
$l_{\infty el} = 135\,nm$ ⊢——⊣

Fig. 4.7. Grain boundary distribution of two silver films ($d_Q = 20nm$) mean lateral crystallite size: $\overline{D}_{lat}$

tribution does neither vary in an unknown way from film to film nor from one area of a film to another. In order to illustrate the grain boundary distribution, Fig. 4.7 shows the lines corresponding to the grain boundaries of two silver films ($d_Q = 20nm$). Both images correspond to an area of about $1.2\mu m \times 0.8\mu m$. They prove that the films show, besides their "individuality", a high degree of reproducibility.

Small Area Micrographs

With higher enlargement individual crystallites can be observed. Thereby, dislocations become visible. The density of observed dislocations equals about 1 per $200nm^2$. If the electron beam is smaller than the crystallites it becomes possible to check the orientation of individual crystallites. A peg count shows that less than 1% of the crystallites have an orientation different from the (111)-texture.

Decoration Technique

Information about the surface structure can be obtained by a special decoration technique. A small amount of gold, in general less than one monolayer, is evaporated onto a silver film or onto other films with low atomic number material. Since gold atoms scatter the electrons more strongly in the transmission electron microscope, even very small gold clusters become visible as

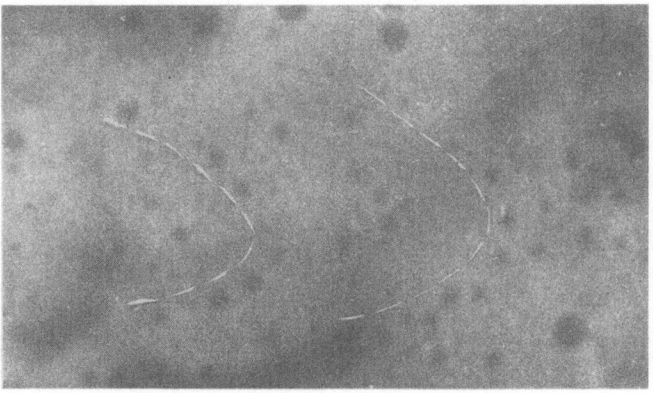

Fig. 4.8. Photograph of a silver film ($d_Q = 20$nm) which is decorated with gold. The gold clusters consist of approximately 10 to 30 gold atoms. White lines are plotted to guide the eye, area: $\approx$30nm×18nm

"little dark patches". This effect can be seen in Fig. 4.8, which shows an area of about 30nm×18nm. The gold atoms are evaporated at 300K. One recognizes that most of the clusters are threaded like a string of pearls. The common theories of crystal growth as developed by Kossel and Stranski help to interpret this photograph (Kossel 1927; Stranski 1928):

The (111)-surface planes of each crystal can have a terraced structure. In the case of a gold decoration at 300K, the terrace edges act as nucleation centers and the larger gold clusters lie along these steps. The mean distance between two following terrace edges can be estimated from such micrographs to be about 5nm.

4.2.2 X-ray Diffraction

In the case of thin films X-ray diffraction techniques are less often used than electron diffraction. This section proves that important comparable and supplementary information can be obtained by this technique. The X-ray diffraction technique often has the advantage of easier application and evaluation. Its disadvantage is in general the small scattering intensity. The scattering volume of a thin film is up to 10^6 times smaller in comparision to the powder diffraction method. However, this can be compensated by longer measuring times and detectors of higher efficiency.

The Diffraction Arrangement

The Bragg-Brentano arrangement is a suitable method to investigate thin films with X-ray diffraction (Glocker 1958). As in the case of the rotating crystal method, a certain gearing rotates the sample at an angle ϑ, while the detector is being rotated at an angle 2ϑ. So the detector registers the beam which is specularly reflected at the atomic planes that lie parallel to the substrate surface. Therefore high intensity can be excected if these atomic planes fulfill Bragg's law

$$2c_{hkl} \cos \vartheta = n\lambda_x, \qquad n = 1, 2, \ldots \tag{4.1}$$

whereby c_{hkl} is the interplanar spacing and λ_x the X-ray wavelength. The measurements are accomplished with the K_α-radiation of copper (weighted mean value λ_x=0.15418nm). A complete diffraction diagram ($10° < 2\vartheta < 160°$) is obtained within one hour(ϑ: angle between the incident beam and the surface normal, 2ϑ: angle between the incident beam and the reflected beam).

Evaluation of the Diffraction Intensities

The diffraction intensities I_{hkl} can be interpreted as a measure of the fraction of crystallites whose orientation $[hkl]$ is perpendicular to the substrate plane. Two conditions must be considered:

- Caused by the aperture of the diffractometer, atomic planes which differ by about $\pm 3°$ from the adjusted orientation also contribute to the diffraction peak.

- In general the planes show a different scattering amplitude due to their packing density.

This influence can approximately be corrected by normalizing the intensities by the relative intensities $I_{hkl,\text{powder}}$ which are obtained for a homogeneous powder. These values are listed in (Powder Diffraction File 1974):

$$I'_{hkl} := I_{hkl}/I_{hkl,\text{powder}}. \tag{4.2}$$

Table 4.1 shows the peak intensities, evaluated this way, for silver films deposited at different substrate temperatures T_D. All films are annealed at about 340K and investigated by diffraction at room temperature. The intensities of

Table 4.1. Normalized relative intensities of the observed diffraction peak as a function of the deposition temperature T_D (normalization see text, homologous peaks are neglected) (Schlemminger, Stark 1986)

T_D/K	(111)	(200)	(220)	(311)	(331)	(420)	(422)
100	1	0.2	0.04	0.35	0.33	-	-
150	1	0.2	0.04	-	-	-	-
200	1	0.05	-	-	-	-	-
225	1	-	-	-	-	-	-
250	1	-	0.04	-	-	-	-
300	1	0.2	0.3	0.46	0.33	0.42	0.54

the (111)-peaks are set to 1 arbitrarily. Homologous peaks are not considered. In general the films are polycrystalline with a majority of crystallites whose [111]-axes are perpendicular to the substrate plane. If the deposition temperature is $T_D = T_D^* = 225K$ (maximum conductivity), the films exhibit a unique (111)-texture.

Figure 4.9 shows the intensity of the (111)-peak I_{111} as a function of the deposition temperature T_D. The values are normalized to the intensity $I_{111}(T_D^* = 225K)$ of a film deposited at a substrate temperature of 225K. The intensity of the (111)-peak reaches a sharp maximum at the deposition

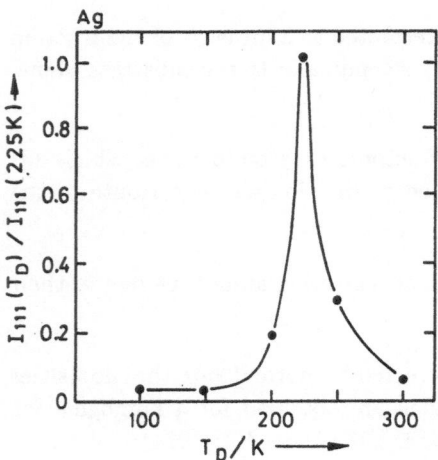

Fig. 4.9. Normalized intensity of the (111)-peak $I_{111}(T_D)/I_{111}(225K)$

temperature T_D^*. A balance of the intensities of the diffraction peaks and the diffuse background intensity proves additionally that at this temperature the distorted fraction of the film shows a significant minimum.

The Half Width of the Diffraction Peaks

Under the entitled assumption of small cubic crystallites, the Scherrer-formula gives a connection between the half width B of the reflection peaks and the thickness of the crystallites d_x (Scherrer 1918; Parrish 1962). Given by the scattering geometry, d_x is the extension of the crystallites perpendicular to the substrate plane:

$$d_x \approx \frac{0.89\lambda}{B_0 \cos \vartheta}.$$ (4.3)

A simple procedure (Kluge, Alexander 1984) permits the elimination of the instrumental broadening of the peak and to determine the true half width. The thickness d_x of the crystals with a [111]-orientation always meets the mean thickness as given by the quartz crystal thickness monitor d_D. This result holds for all deposition temperatures between 100K and 300K and not only for T_D^*. However, the thicknesses d_x of crystallites with other orientations appearing for $T_D^* \neq T_D$ are always significantly smaller as derived from the Scherrer-formula. The described results are compiled in Fig. 4.10. A detailed

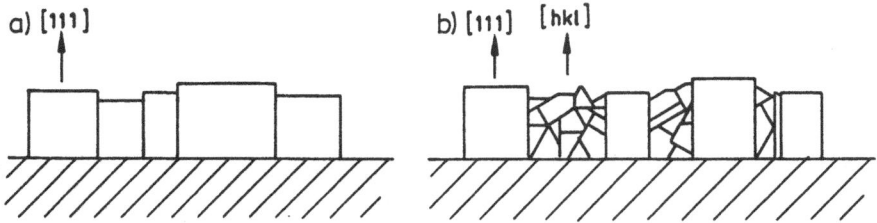

Fig. 4.10. Crystalline structure of thin evaporated silver films on glass supports after the annealing procedure with a change of the deposition temperature. a) deposition temperature: $T_D = T_D^*$, b) deposition temperature: $T_D \neq T_D^*$

analysis shows that the lattice compression or dilatation also reaches a minimum at T_D^* (Schlemminger 1982). The described behaviour is not limited to silver films, but seems to be quite universal (Table 4.2). For the specified metals a unique texture was obtained at the deposition T_D^* which guarantees a maximum dc-conductivity (Schlemminger 1982; Schlemminger 1989).

Fine-Structure of the Diffraction Peaks

Under suitable conditions, the diffraction peaks obtained from smooth thin films show secondary maxima (fringes), as known from optical gratings. The instrumental broadening and the inhomogenity of the film thickness give a special shape. Croce et al. were the first who tried to obtain information about the surface roughness from the fringes (Croce, Devant, Verhaege 1965).

Table 4.2. Deposition temperature T_D which lead to an unique texture

metal	structure	T_D^*	texture
Au	fc	250	[111]
Ag	fcc	225	[111]
Cu	fcc	310	[111]
In	tetragonal	110	[011]
Mg	hcp	140	[001]

Fischer and Wißmann provided a model which permits a quantitative evaluation of the film thickness inhomogenities (Fischer, Wißmann 1982). This procedure is used here:

Unlike the X-ray diffraction of solids, in the case of a thin film only a small number of atomic planes N_{hkl} contribute to the diffraction signal $I_{hkl}(N_{hkl}, \varphi)$. For a textured or monocrystalline silver film with thickness $d = 20\text{nm}$, $N_{hkl} \approx 85$. Therefore a perfectly orientated film gives rise to a diffraction pattern of the form:

$$I_{hkl}(N_{hkl}, \varphi) = N_{hkl}^2 u^2 \frac{\sin \varphi}{\varphi} \qquad (4.4)$$

with

$$\varphi = 2\pi c_{hkl} \Delta\vartheta \lambda_x \cos \vartheta.$$

In this equation $\Delta\vartheta$ is the deviation from the Bragg-angle ϑ, c_{hkl} is the interplanar spacing, u the scattering amplitude of one plane, and λ_x the wavelength of the applied radiation. A real film deviates from this situation in a certain way. First of all, it is obvious that the crystallites have different heights due to the growth process. Here it shall be assumed that the outer boundaries of the grains are parallel planes. The discrete function $f(N_{hkl})$ describes the distribution of the crystallite heights $N_{hkl} c_{hkl}$ (see Fig. 4.11). The distribution $f(N_{hkl})$ is approximated by a Gaussian-distribution function, with a mean

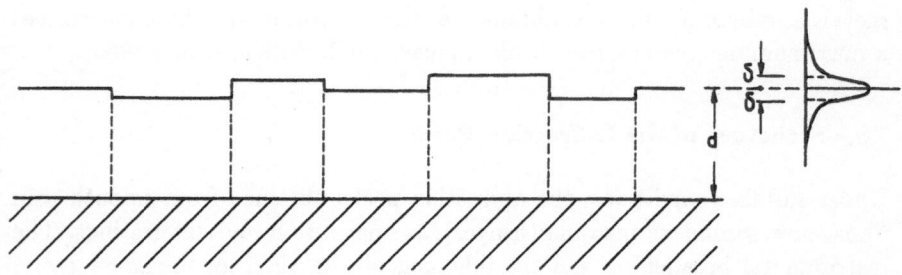

Fig. 4.11. Distribution of the crystallite heights of a textured or monocrystalline film

thickness d_M and a mean deviation δ. The scattering intensities of all crystallites are now incoherently superposed:

$$I_{hkl}(\varphi) = \sum_{N_{hkl}} f(N_{hkl}) I_{hkl}(N_{hkl}, \varphi). \qquad (4.5)$$

The theoretical result of this approach can be seen in Fig. 4.12 for different values of the relative mean deviation $r := \delta/d$ (Häupl, Wißmann 1984). With

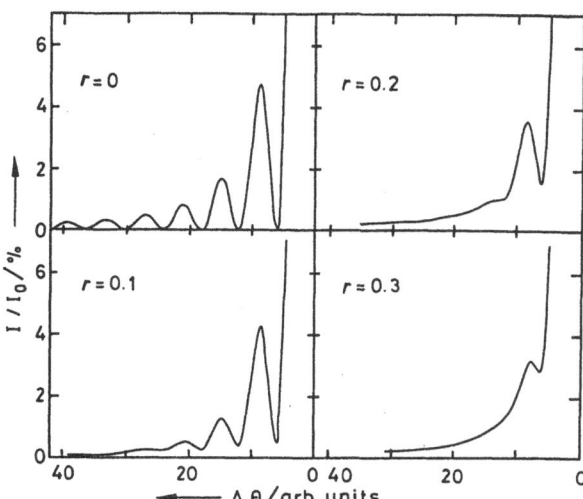

Fig. 4.12. Calculated fine-structure of the diffraction peak, due to different relative mean deviations $r := \delta/d$ of the crystallite heights, $r = 0, 0.1, 0.2, 0.3$ (Häupl, Wißmann 1984)

an increasing inhomogenity of the film thickness the fringes are smeared out. Figure 4.13 shows the measured fine-structure of the (111)-peak of a 20nm thick silver film. The film was prepared under the above mentioned conditions to obtain maximum conductivity. Diffraction peaks of films evaporated under different conditions have never shown a fine-structure. A procedure to extract the value r from the measurement independently from the instrumental broadening of the central peak was given by Häupl and Wißmann (Häupl, Wißmann 1984). An evaluation along this line shows that the mean relative variation of the crystallite heigths r over the whole film area is lower than 5%. This is a surprising result, since such values could otherwise only be obtained from monocrystalline films (Häupl, Wißmann 1984).

The fringes contain additional important information. Using Bragg's law for the upper and lower film boundary the film thickness d_F can be calculated from the mean angular distance $\overline{\Delta\vartheta}$ between the fringes.

Though this thickness measurement only works between about 10nm and 30nm, it is relevant since it is not based on the usual assumption that a certain quality of the film equals the bulk value. Here these values are in good agreement with the data obtained from the quartz crystalmonitor.

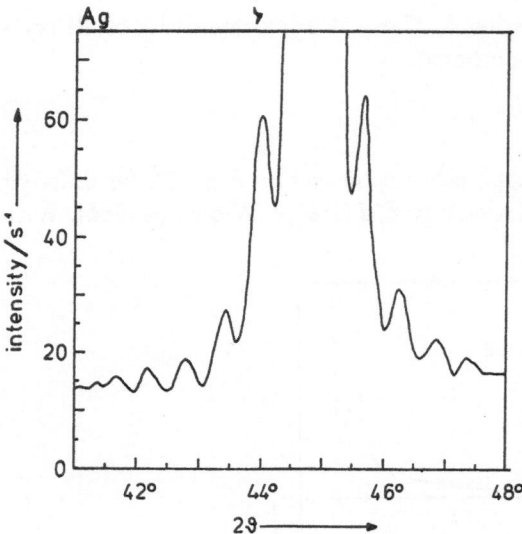

Fig. 4.13. Fine-structure of the (111)-peak of a 20nm thick silver film (Schlemminger 1989)

4.2.3 Scanning Tunnelling Microscopy

Scanning tunnelling microscopy (STM) enables one to visualize the surface structure of the films at an atomic scale. Therefore STM studies are necessary in this context. The micrographs shown and described here were made with a STM developed by Besocke (Besocke 1987). Figure 4.14 shows a y-modulation image of a silver film which was prepared as decribed above. The tip voltage is approximately +0.1V. The STM was operated in the constant current mode ($I = 10^{-9}$A). The shown area has a size of $\approx$100nm$\times$75nm. Several crystallites can be seen. On the rather extended crystallite in the background the terraces and terrace edges are well resolved. The height of the terrace edges are 0.25nm $\pm$0.05nm which is consistent with the interplanar spacing in [111]-direction. All crystallites exhibit the same terraced surface structure. It can be made visible by rotating the sample. The terraces are in general atomically smooth. The mean width of the terraces perpendicular to the edges is about 5nm, while the width of the terraces along the edges equals the lateral crystallite size. Figure 4.15 shows an STM-micrograph with higher magnification. The plotted area equals about 50nm$\times$30nm. The mapping visualizes an individual crystallite and the proximity of a grain boundary. At the grain boundary a steep incline leads downward to the next crystal. The crystallite heights differ by about 1nm, which corresponds to the value obtained from the X-ray diffraction fine-structure. Besides the silver films Au-, Cu-, and In-films were investigated. The last two metals form non- or semiconducting oxide layers at the air, which make an imaging by the STM nearly impossible. However, a

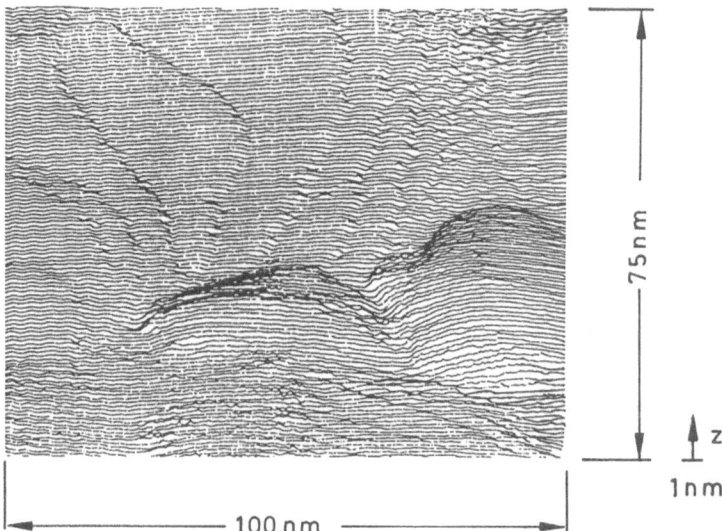

75 nm

z

1 nm

100 nm

Fig. 4.14. STM-micrograph of the surface of a silver film (d_Q=20nm), area: ≈100nm×75nm

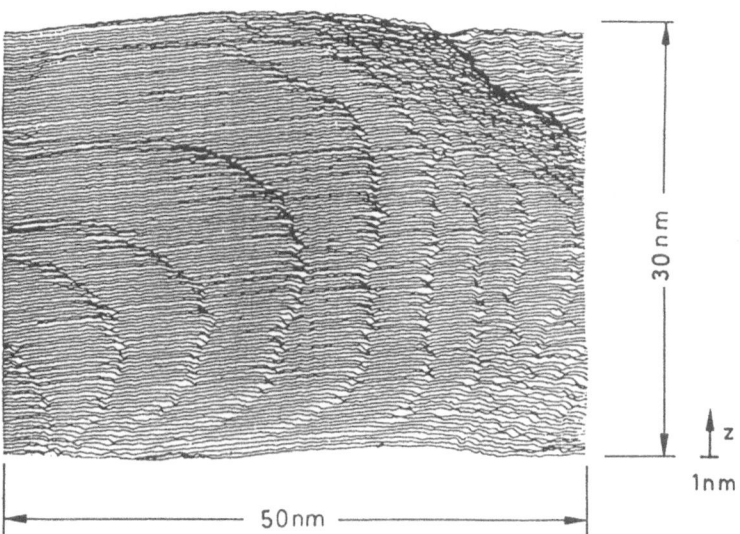

30 nm

z

1 nm

50nm

Fig. 4.15. STM-micrograph of an individual crystal and the proximity of a grain boundary, area: ≈50nm×30nm

coverage of two monolayers of gold under ultra high vacuum conditions at a suitable temperature can avoid the oxidation. It can be assumed that the low gold coverage contours the copper and indium films. Pictures of these films exhibit the same structural features as the thin silver films. Only the pure gold films show a striking structural difference. The terrace edges of these films consist of rather straight parts which always create angles of 120° or 60° to each other.

4.2.4 Electrical Conductivity

The measurement of the electrical conductivity has already been used to find the optimum evaporation conditions as described in the first part of this chapter. Here the electrical properties of the films are described independently of the controversial thin film conductivity models as much as possible.

DC-resistivity Measurements

In order to measure the film resistance, contact films with a thickness of 100nm to 200nm are evaporated onto the glass supports in a leading step. A suitable arrangement of the evaporation source, the mask, and the sample serve for wedge-shaped borders of the contacts. Gold wires (Ø=0.05mm) are adhered to the contact-films with a solution of silver particles in an organic solvent. The solvent vapourizes during a following heat treatment and the remaining silver provides a good mechanical stability and a low contact resistance in a wide temperature range. Four (or sometimes six) wires serve as separated current and voltage contacts. A displaceable mask permits the evaporation of up to 20 films without breaking the vacuum side by side on one substrate. A 'scraper' disconnects the previously prepared and measured film in each case. Therefore for all films the same contacts and wires can be used. The standard resistance measuring circuit is shown in Fig. 4.16. It enables one to measure the film resistance R with a four terminal arrangement and to register sensitive changes of the film resistance ΔR. Since the resistances

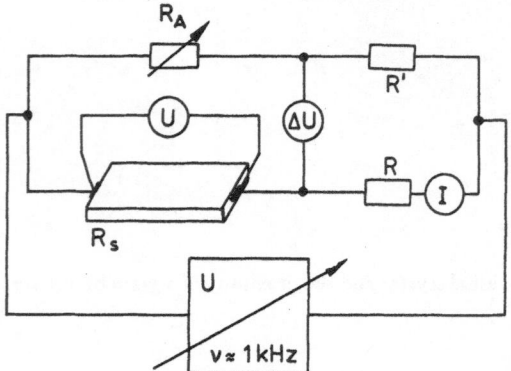

Fig. 4.16. Measuring circuit I

R and R' are much higher than R_s and R_A and the voltage-meters U and ΔU have high ohmic inputs, the current I is constant for changes of the film resistance in a limited range. Therefore the voltages U and ΔU are directly proportional to R and ΔR. The circuit is supplied with an alternating current

($\nu \approx 1\mathrm{kHz}$) because it is technically easier to detect small ac-voltages than dc-voltages. Additionally, the influence of thermo-voltages is eliminated. In a physical sense an alternating current of about 1kHz can be regarded as dc-case. The voltages U and ΔU are measured with a lock-in amplifier which is used as a narrow band voltage-meter. In the case of small resistance changes, a suitable signal to noise ratio is also reached. The current density is about $5\mathrm{A/mm^2}$. The ohmic behaviour of the films is tested by varying the current density by more than three orders of magnitude. In order to minimize the thermal drift and noise the circuit is mounted in a thick-walled metal case directly onto the ultra high vacuum current feed-through. In order to obtain a stable signal, a renunciation of plugs and switches in the circuit is of high importance.

In general the output of the lock-in amplifier is directly recorded as a function of temperature, film thickness, time, etc.

Thickness Dependence of the Film Resistivity

In order to get information about the electrical properties of the thin "sensor-films", the thickness and temperature dependence was measured. If Matthiessen's-rule is valid it can be assumed that the film resistivity might be divided into a thickness independent fraction A and a thickness dependent fraction $B(d)$.

Some plausible assumptions can be made. The scattering centres which cause the thickness dependent fraction of the resistivity are located at the inner and outer boundaries of the crystallites (see Fig. 4.17, hatched areas). They have a certain and thickness independent two-dimensional density at

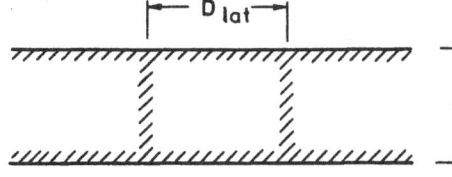

Fig. 4.17. Schematic distribution of the defects (given by the hatched regions) which cause a thickness dependent fraction of the film resistivity

the boundaries. In the case of textured films as used here the mean lateral dimension of the grains $\overline{D}_{\mathrm{lat}}$ is proportional to the film thickness d. Then the three-dimensional density of the scattering centres with respect to the whole volume of the film is proportional to the reciprocal film thickness. Therefore at first approximation it can be expected that the thickness dependent part of the film resistivity ϱ is proportional to d^{-1}, too.

Therefore it is sensible to use a linearized plot of $\varrho(d) \times d$ versus d as shown in Fig. 4.18. Obviously there exists a linear dependence for films in the thickness range between 20nm and 200nm. Below 17nm the films exhibit

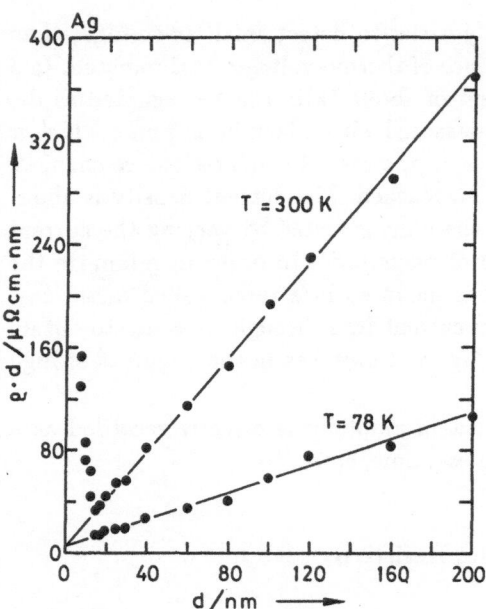

holes which cause significant deviations. These films must be excluded from an evaluation. Indeed a simple linear dependence remains.

At a fixed temperature such a curve contains only two parameters. In the case of silver:

$A = 1.84 \mu\Omega\text{cm}$ at 300K and $A = 0.51 \mu\Omega\text{cm}$ at 78K

$B = 5.3 \mu\Omega\text{cm·nm}$ at 300K and 78K.

It is worthless to evaluate these data in a model with more than two free parameters. Here the aim is only to estimate the maximum influence of the surface- and grain boundary scattering. In the framework of Fuchs' model using Sondheimer's approximation, A can be identified with ϱ_∞ and under the assumption that $(\varrho \cdot l)_\infty = (\varrho \cdot l)_{\text{poly}} = 843 \Omega\text{nm}^2$ the mean specularity parameter $\bar{p} \approx 0.83$ can be calculated. This value can be interpreted as a lower limit, since at the same time grain boundary scattering causes a thickness dependent contribution. With the Mayadas and Shatzkes approach the maximum influence of this effect can be estimated. With the help of a common approximation (Wißmann 1975). The value A can be identified with ϱ_B, which differs in its meaning from ϱ_∞ as explained above (Sect. 2.3). Under the assumption that $(\varrho \cdot l)_B = (\varrho \cdot l)_{\text{poly}} = 843 \Omega\text{nm}^2$ and with the knowledge that $\overline{D}_{\text{lat}} \approx 1.5 \cdot d$, the upper limit of the grain boundary reflectivity as defined in the model can be calculated: $R_g \leq 0.06$. These values qualify the films as very suitable to observe the influence of surface processes on the electrical resistivity. Gold, copper and indium films prepared as described above show a similar behaviour.

The Temperature Dependence of the Resistivity

During the annealing process the film resistivity decreases irreversibly. Thereafter the film resistivity always changes reversibly and linearly with the temperature between 20K and T_A. For continuous films the absolute temperature coefficient of the resistivity (ATCR) $\beta := \partial\varrho/\partial T$ does not depend on the film thickness and meets the value of polycrystalline silver $\beta_{\text{poly}} = 0.0061\,\mu\Omega\text{cm/K}$ within the measurement accuracy. This result sup-

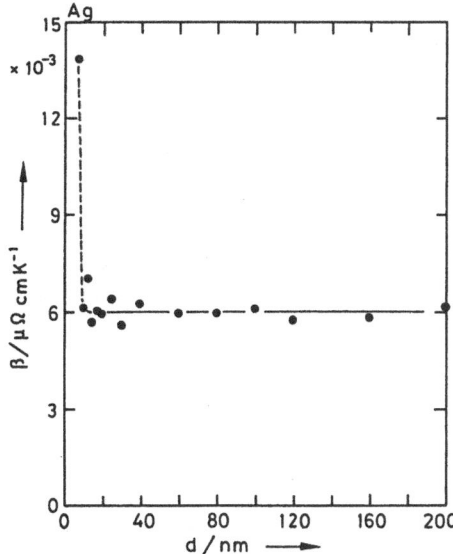

Fig. 4.19. The absolute temperature coefficient of the resistivity (ATCR) $\beta = \partial\varrho/\partial T$ versus film thickness d

ports the validity of Matthiessen's rule and permits a very helpful technique to study temperature dependent surface phenomena.

Temperature Dependent Surface Processes

In order to investigate surface processes as a function of the substrate temperature the influence of the temperature coefficient on the conductivity must be eliminated. In case A two almost equal films (deviation in resistivity $\leq$ 1%) are prepared. Both films act as parts of a symmetric Wheatstone-bridge (see Fig. 4.20) In this measuring circuit R and R' are also much higher than R_s and R'_s in order to determine the current I. With the help of the potentiometer P the bridge can be adjusted. Since both films can differ not only in the temperature dependent part of the resistance but also in the temperature independent defect-induced contribution, an additional adjustment, realized as a zero point suppression at the lock-in amplifier is necessary (see Fig. 4.20). All other components of the circuit are the same as described above. Separate

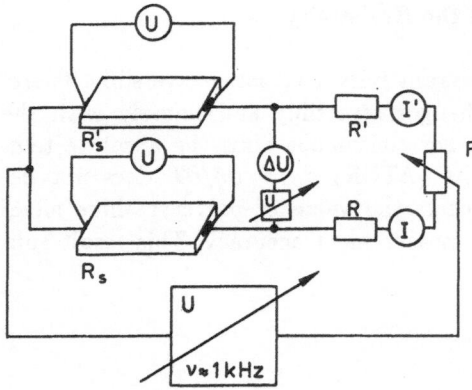

Fig. 4.20. Measuring circuit II, case A: $(d \approx d')$ $R_S \approx R'_S \ll R \approx R'$, case B: $(2d \approx d')$ $R_S \approx 2R'_S \ll R \approx 2R'$

shutters in the ultra high vacuum chamber permit the evaporation of adma-terial alternatively onto one or both films. In the case of gases a selective exposure is not easy to realize. Therefore the measuring circuit is used under other conditions (case B). Two films with defined different thicknesses are prepared. In the thickness region of interest the resistance ratio R_s/R'_s equals nearly the reciprocal thickness ratio d'/d and the absolute temperature coef-ficient is not temperature dependent, as shown above. The bridge is *barely* adjusted in a wide temperature range if $R_s/R'_s \approx I'/I \approx R/R'$. The tempera-ture independent resistances which differ slightly are compensated for with the zero point suppression at the lock-in amplifier. Due to the imperfect adjust-ment it is necessary to determine these deviations as a function of the sample temperature in a previous experiment with the "identical" films. The resis-tance changes caused by a surface influence are approximately proportional to d^{-2} and d'^{-2}. This meets the theoretical prediction and is experimentally proven (see Sect. 5.1). Therefore the voltage-drops due to surface effects are nearly proportional to the reciprocal thicknesses d^{-1}, d'^{-1}. It is reliable to use the film thicknesses $d = 20$nm and $d' = 40$nm since the suppression of the temperature dependence works sufficiently and the signal is only reduced by a factor 2. The sensitivity and accuracy of the dc-resistance measurement are limited by undesirable temperature variations of the sample. In general, an accuracy of about 10^{-4} can be reached.

5. Studies of Surface and Growth Processes

In this chapter examples will be given to outline the power of the tool "conductivity measurements" to study surface- and growth processes. Results obtained by other authors will only be included as far as their films are characterized well enough in the given context.

5.1 Adsorption and Desorption

The starting point is the reproduction of Lucas' experiment (Lucas 1964) explained above under well defined conditions. Here a silver base film of the described type is used. Additional silver is evaporated with a rate of 10 monolayers per second under ultra-high vacuum conditions. Figure 5.1 shows the resulting change of the resistivity $\Delta\varrho$ versus coverage θ in the range between 0 and 0.5[1]. If the substrate temperature is high enough ($T = 350$K), the resistivity hardly varies with coverage. The temperature $T = 350$K equals the annealing temperature T_A. Thus it is evident that the surface structure cannot change significantly when the same material is slowly added. The impinging adparticles diffuse across the surface until they are incorporated at the terrace edges. At a substrate temperature of 10K the adatoms have nearly no mobility (Schumacher, Stark 1982). They cover the terraces statistically unordered. So at low coverages every adatom acts as an additional scattering centre for the conduction electrons striking the surface. It is sensible to compare the influence of a monomer on a perfect (111)-terrace with the resistivity increase caused by a bulk point defect. Their contribution to the resistivity is 1.5 to $2.3\mu\Omega$cm / at.% for interstitials (Potter, Dexter 1957) and 1.3 to 1.5 $\mu\Omega$cm / at.% for vacancies (Henderson 1972) in noble metals. From the initial slope the following can be deduced:

– Ag-monomer on Ag(111): (1.3 ± 0.1) $\mu\Omega$cm / at.%

– Au-monomer on Au(111): (1.1 ± 0.1) $\mu\Omega$cm / at.%, see (Pariset, Chauvineau 1978)

---

[1]In the case of metals the coverage θ referes to a complete monolayer as deduced from the quartz thin film monitor. Here and in the following Figures at first position the admaterial and behind the slash the metal of the base film is given. The texture of the base film is noted in Table 4.2.

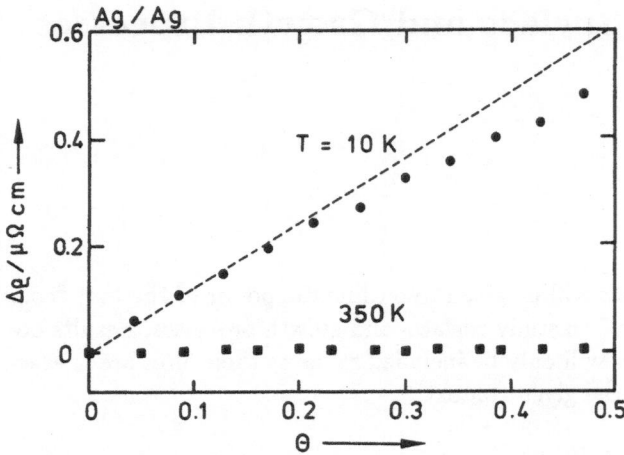

Fig. 5.1. Change of the resistivity $\Delta\varrho$ of a silver film during coverage θ with additional silver at different substrate temperatures T, base film thickness $d_1 = 20$nm, dashed line: initial slope, measuring circuit: I

  – Cu-monomer on Cu(111): (0.6 ± 0.2) $\mu\Omega$cm / at.%, see (Roth, Schumacher, Stark 1992).

Persson has related the resistivity increase $\Delta\varrho$ and the total cross section for diffus scattering Σ, respectively, to the live time of the frustrated translation vibration $\tau_{\text{e-h}}$ of the adparticle (Persson, Schumacher, Otto 1991), (Chap. 3). Considering a Newns Anderson type of model of chemisorption he has derived a relation to determine this quantity (Persson 1991). With the help of this value the cross-section can be calculated (Γ is the width of the adsorbate induced density of states ρ_a):

$$\Sigma \approx \frac{32}{3} \frac{E_F}{\hbar v_F} \frac{<\sin^2\theta>}{n} \Gamma \rho_a(E_F). \tag{5.1}$$

From the weak change in the work function for single silver adatoms on a Ag(111) surface one can deduce that the Ag atom level (5s) forms a half filled resonance near or centered at E_F. Therefore in the present case $\Gamma\rho_a(E_F) \approx 2/\pi$ and (5.1) becomes:

$$\Sigma \approx \frac{64}{3\pi} \frac{E_F}{\hbar v_F} \frac{<\sin^2\theta>}{n}. \tag{5.2}$$

For silver $E_F = 5.5$eV, $v_F = 1.4 \times 10^6$m/s and $n = 5.86 \times 10^{28}$m^{-3}. The quantity $<\sin^2\theta>$ considers the overlapp between the metal electron wave function and the adatome-state. Its value is approximately 0.2 [For details see: Persson (1991)]. The result $\Sigma_{\text{theo}} = 0.134$nm^2 is in very good aggreement with the experimental value $\Sigma_{\text{exp}} = 0.146$nm^2 obtained from Fig. 5.2 with the help of (3.2).

For a better understanding and in order to avoid fundamental errors it is necessary to check the thickness dependence of the absolute resistivity change. The existing models concerning the classical size effect predict a proportionality to d^{-1} as a first approach. Such a dependence has been seen for example, for CO adsorbed on Ni films by Wißmann (Wißmann 1975) and for gold on gold films at low temperatures by Fischer (Fischer 1980). The chosen doublelogarithmic plots wipe out the high uncertainty of the measurements. The use of more suitable and sufficiently constant preparation conditions, see above, leads to the results compiled in Fig. 5.2. Here the initial slope $\partial\varrho/\partial\theta$ for the condensation of additional silver on silver films at 10K is plotted versus the reciprocal film thickness. Indeed the linearity is clearly fulfilled down to a film thickness of 12.5nm. This linearity proves that it is justified to interpret the described phenomena as surface effects.

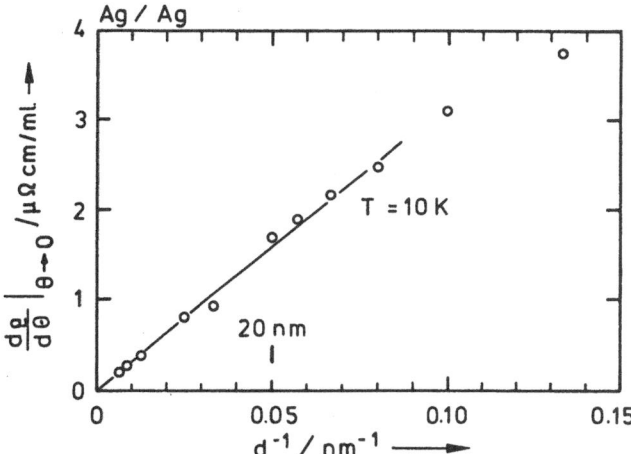

Fig. 5.2. Initial slope $\partial\varrho/\partial\theta$ of the resistivity for the condensation of silver on silver films at a substrate temperature of 10K versus the reciprocal film thickness d^{-1}

Foreign Admetals

The evaporation of foreign metals onto smooth thin films always results in a resistance increase. Pariset and Chauvineau have compared the covered thin film with a diluted alloy (Pariset, Chauvineau 1978). The resistivity behaviour of diluted alloys was extensively investigated by Norbury and Linde (Norbury 1921; Linde 1931, 1932). Empirically they found the relation that the resistivity increase per percent foreign atoms c is proportional to the second power of the valence difference between the solvent and the impurity $(\Delta Z)^2$.

The trail to transfer this explanation by Pariset and Chauvineau was less satisfying (Pariset, Chauvineau 1978). Their data $\Delta\varrho/c$ vs. $(\Delta Z)^2$ ly on a strait line which does not meet the origin but shows a significant and unexplained intersection and has a much smaller slope than for bulk material. In Persson's model the resistivity increase is proportional to the density of

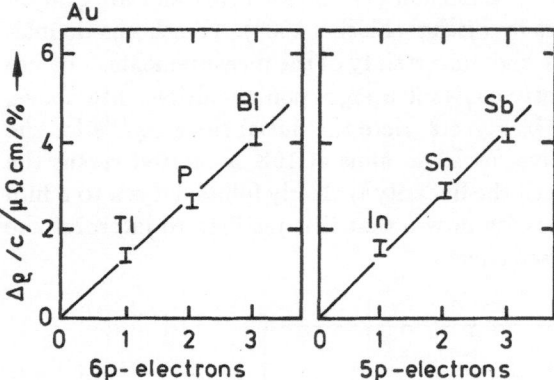

Fig. 5.3. Additional resistivity per percent of impurity atoms versus the number of valence electrons. Data taken from (Pariset, Chauvineau 1978)

the adsorbate induced state ρ_a at E_F (5.1). However, this value is directly related to the number of valence electrons of the adatoms. Figure 5.3 shows an evaluation of the data taken from Pariset and Chauvineau in this sense. The clear proportionality is a further support of Persson's approach.

Gas-coverage

The influence of gas-coverage has been studied for a rather long time (Wißmann 1975; Dayal, Finzel, Wißmann 1987; Wedler 1987). In the case of sufficiently smooth metal films gas-adsorption always leads to a resistance increase (Wißmann 1975; Dayal, Finzel, Wißmann 1987; Geus 1971 and references therein). Common features are the linearity of the initial slope and a saturation for higher coverages. The absolute values and the special form of the curve depend on the gas species. Sometimes a resistivity decrease was observed for hydrogen exposures (Suhrmann 1957). However, insufficient vacuum conditions permit the explanation that those films were covered with oxygen or a metal oxide. Therefore the hydrogen exposure can have led to an uncontrolled chemical reduction process at the surface. Recently Holzapfel et al. (Holzapfel et al. 1990b) have shown a clear correlation between the height of the resistivity increase and electronic structure of the adsorbate in the case of several hydrocarbons. The initial slope and the saturation value decrease with increasing energetic distance between the Fermi-level of the substrate (silver) and the lowest unoccupied molecular orbital of the adsorbate. This dependence fits well into Persson's model which predicts that the initial slope shoud be proportional to the local electron density of the adsorbate induced state at E_F, see above (Persson, Schumacher, Otto 1991; Persson 1991; Grabhorn et al. 1991).

Reproducibility and Desorption

Generally, a reproducibility within several percent can easily be achieved for the same film for adsorbed gases or condensed metal-atoms of the same material. The film is heated until the gas desorbs or the atomic scale roughness is annealed. A new coverage with gas or metal atoms leads to the same characteristic resistivity behaviour. The endeavour towards optimum and constant evaporation conditions has led to a difference from film to film under 10% on the same substrate and below 20% on different glass slides.

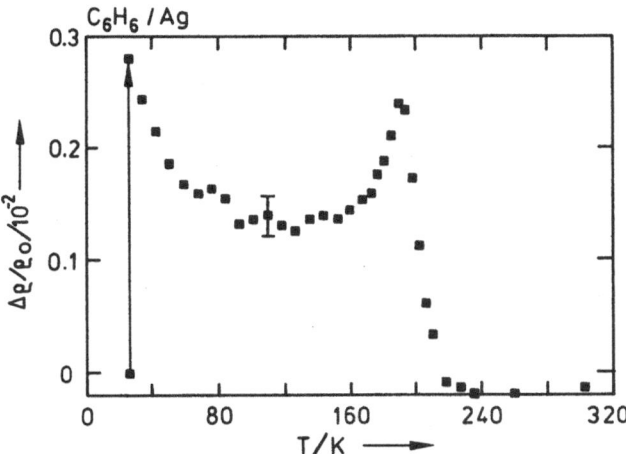

Fig. 5.4. Relative change of the resistivity $\Delta\varrho/\varrho_0$ of a 20nm thick silver film during the adsorption (coverage: $\theta \approx 0.1$) and desorption of C_6H_6. Measuring circuit II (Tonscheidt 1990)

More interesting than the strictly speaking obvious reproducibility is the possibility to monitor both the adsorption and the desorption of a gas cyclically. The first resistivity measurement which has shown such a process is represented in Fig. 5.4. At a sample temperature of 30K approximately one tenth of a monolayer[2] of benzene (C_6H_6) is adsorbed on two silver films which are of different thickness and parts of the measuring equipment II. This leads to a resistivity increase of nearly 0.3% for the 20nm thick film (see arrow, Fig. 5.4). The small coverage ensures that the C_6H_6 molecules are directly bound to the silver surface. The comparatively small resistivity change causes a large error bar, as also shown in Fig. 5.4. After the adsorption the films are heated with a rate of 0.1K/s and at the desorption temperature of benzene

---

[2]In the case of gas adsorption the coverage θ refers to a complete monolayer as deduced from thermal desorption spectroscopy measurements.

($\approx$ 200K) a sharp decrease of the additional resistivity occurs which leads back to the starting value within the measuring accuracy. During the heating between adsorption and desorption the resistivity does not remain constant. The decrease between 30K and 80K might probably be caused by agglomeration processes whereas the increase between 160K and 200K might be attributed to a rising scattering cross-section of the vibrating molecules. The whole process can be repeated with the same result and monitored by the dc-resistivity measurement.

Catalytic Reaction

The oxidation of ethylene (C_2H_4) on the surface of a silver film is an example, taken from the thesis of Wittmann (Wittmann 1984) which shows that catalytic reactions can also be observed by surface scattering experiments with conduction electrons:

The adsorption of O_2 on a 30nm thick silver film at room temperature causes a significant resistivity increase. During the following exposure to C_2H_4 the resistivity decreases again. Mass-spectroscopy shows that at the same time CO_2 and H_2O molecules leave the film surface proving the catalytic reaction. The formation of ethylene oxide cannot be expected at room temperature.

If the experiment is repeated at 77K the exposure to C_2H_4 has no influence on the film resistance and no reaction products can be observed. Since the surface structure is not clearly defined and the surface roughness at atomic scale is not systematically varied, this experiment can only hold as an example that it is in principle possible to observe surface reactions using dc-resistivity measurements.

5.2 Diffusion Processes

Since a lot of questions are unsolved and under discussion in the subject of surface diffusion, it is reasonable to use the dc-resistivity measurements in this context.

Surface Self-diffusion

The metal films described above can act as suitable substrates because they exhibit a rather unique surface structure which can be described in the terrace-ledge-kink model (Stranski 1928; Kossel 1927; Burton, Cabrera, Frank 1951). Figure 5.5 shows a sphere model of the film surface. It visualizes possible adatom positions and defects. The possibility to monitor surface self-diffusion processes with the help of dc-resistivity measurements is based on the fact that an adatom causes different resistivity changes depending on its position on a terraced surface (Chauvineau 1980):

Fig. 5.5. Sphere model of the film surface (fcc(111)-plane) with different defects and adatom positions (see text)

- An isolated adatom on the perfect terrace causes a resistivity increase (position 1, Fig. 5.5).

- If it fills a surface vacancy in the terrace plane (position 2), it effects a resistivity decrease.

- The contribution of the single atom is significantly smaller if it is taken up at the terrace edge (position 3).

- A resistivity decrease can be expected, if it fills a vacancy in the terrace edge (position 4).

- An adatom which is incorporated at a kink (position 5) does not have an influence on the film resistivity in the first approximation.

In principle it should be possible to get information about surface self-diffusion processes by measuring the film resistivity during a small coverage and a following heat treatment. This method is comparable with the technique to study surface self-diffusion by work function measurements with the Kelvin-method (Hölzl, Schulte 1979). There dipole-moments are attributed to each adatom concerning its coordination number. The sum of the dipole-moments can be connected with a change in work function. In both cases a macroscopic quantity contains information about different microscopic processes. Therefore an elaborate model of surface diffusion is necessary. Schrammen and Hölzl (Schrammen, Hölzl 1983) have used computer simulations to determine the activation energies of the diffusion process in the case of Ni on Ni(100).

A first diffusion experiment of this type on the base of a resistivity measurement was done by Chauvineau (Chauvineau 1980) with gold on a gold film. Figure 5.6 shows a very similar result for silver on a silver film. At a temperature of approximately 50K about 10^{-1} of a monolayer of silver is evaporated onto a silver film. As expected a resistivity increase results. The additional resistance disappears completely when the film is heated continuously to 350K. Between $T \approx 50K$ and $T \approx 60K$ the additional resistivity remains

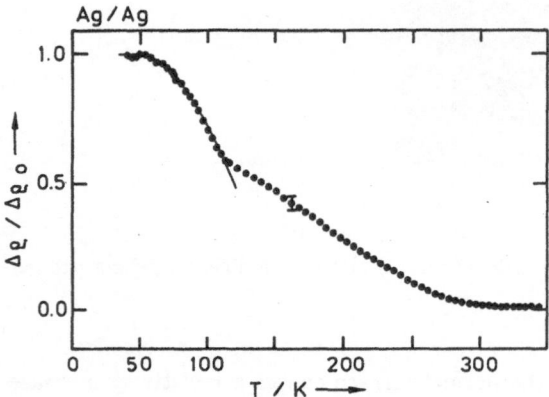

Fig. 5.6. Normalized resistivity increase of a 20nm thick silver film due to a coverage of $\theta \approx 2 \times 10^{-2}$ and the resistivity decrease caused by a following heat treatment (Tonscheidt 1990)

constant. Then the annealing of the surface defects obviously occurs stepwise. Following Chauvineau's interpretation the fist step between $T \approx 60K$ and $T \approx 100K$ can be attributed to the disappearance of monomers. The low coverage ensures that isolated adatoms reach the terrace edges and will be caught there. The following resistivity decrease is dominated by the annealing at the terrace edges. Above $T \approx 330K$ the original surface structure seems to be recovered, since the additional resistivity completely dissappears. Self-diffusion experiments with other metal films qualitatively deliver the same results:

Ag on Ag, $\theta \approx 5 \times 10^{-2}$ (Schumacher, Stark 1983),

Au on Au, $\theta \approx 1 \times 10^{-2}$ (Chauvineau 1980),

Bi on Bi, $\theta \approx 5 \times 10^{-4}$ (Chauvineau 1980),

Cu on Cu, $\theta \approx 1 \times 10^{-2}$ (Roth, Stark 1990).

The first part of the curve can be evaluated with the help of a simple diffusion model. Details and difficulties are explained elsewere (Roth, Schumacher, Stark 1992). Here only a short description will be given:

A thermally activated two-dimensional diffusion process of monomers is assumed (Ehrlich, Stolt 1980):

$$D = D_0 \exp\left(-\frac{E_0}{k_{\mathrm{B}}T}\right) \tag{5.3}$$

D : diffusion coefficient, D_0 : diffusion constant, E_0 : activation energy.

This description follows the ideas of (Burton, Cabrera, Frank 1951).

A Surface Model

It is sufficient to describe the process on one terrace plane (see Fig. 5.7). For simplification its edges are positioned at $x = -1/2$ and at $x = +1/2$ so that the terrace width s is defined as the unit length. The statistically equipartition of the adatoms at the time $t = 0$ is described by a homogeneous surface density n_0. At this time the temperature is T_0. During the following diffusion

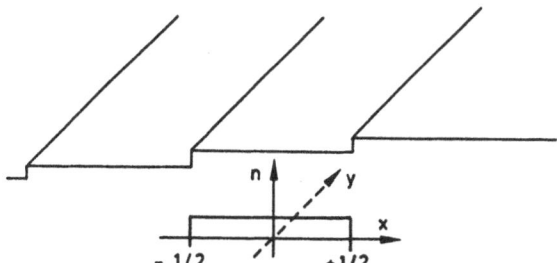

Fig. 5.7. Simple model of the terraced surface

process the edges act as ideal traps for the monomers. Then the depopulation of the terrace can be described by the equation of one-dimensional diffusion (Crank 1967):

$$\frac{\partial n(x,t)}{\partial t} = D(T(t))\frac{\partial^2 n(x,t)}{\partial x^2}. \tag{5.4}$$

There $n(x,t)$ is the local density of monomers as a function of time and the one-dimensional space coordinate x. An approximate solution of this equation for long diffusion times was given by Stark (Stark 1987). This cannot be applied here because for long diffusion times the resistance behaviour is dominated by the processes at the edges. The general solution can easily be derived introducing a suitable time scale τ, which leads to a simpler equation:

$$\frac{\partial n(x,\tau)}{\partial \tau} = \frac{\partial^2 n(x,\tau)}{\partial x^2}, \qquad \tau := \int_0^t D(\vartheta)d\vartheta. \tag{5.5}$$

If the sample is heated linearly $T = T_0 + r_A t$ with the rate r_A starting from the deposition temperature T_0, it results:

$$\tau = D_0 \left(\frac{1}{s}\right)^2 \int_0^t \exp\left(\frac{-E_0}{k_B(T_0 + r_A\vartheta)}\right) d\vartheta. \qquad (5.6)$$

The boundary conditions are:

$$n(x,\tau) = n_0 \quad \text{for} \quad |x| < 1/2 \quad \text{and} \quad \tau = 0, t = 0 \qquad (5.7)$$
$$n(x,\tau) = 0 \quad \text{for} \quad |x| \geq 1/2 \quad \text{and} \quad \tau \geq 0, t \geq 0.$$

The solution is:

$$\frac{n(x,\tau)}{n_0} = 1 - \sum_{m=0}^{\infty}(-1)^m \left(\text{erfc}\frac{(m+\frac{1}{2}) - x}{2\sqrt{\tau}} - \text{erfc}\frac{(m+\frac{1}{2}) + x}{2\sqrt{\tau}}\right) \qquad (5.8)$$

with $\quad \text{erfc}(z) := 1 - \frac{2}{\sqrt{\pi}} \int_0^z \exp(-y^2) dy.$

The additional resistivity is connected with the mean surface density of ad-particles. A proportionality between $\Delta\varrho$ and $\bar{n}$ can be assumed for small coverages.

$$\frac{\Delta\varrho(\tau)}{\Delta\varrho_0} \approx \frac{\bar{n}(\tau)}{n_0} = \int_{-1/2}^{1/2} \frac{n(x,\tau)}{n_0} dx \qquad (5.9)$$

$$= 1 - 4\sqrt{\tau} \sum_{m=0}^{\infty}(-1)^m \left(\text{ierfc}(\frac{m}{2\sqrt{\tau}}) - \text{ierfc}(\frac{m+1}{2\sqrt{\tau}})\right)$$

with $\quad \text{ierfc}(z) := \int_z^{\infty} \text{erfc}(y) dy.$

The parameters E_0 and D_0 can be obtained by a numerical evaluation of (5.9) and a fit procedure (Roth 1989). In the case of Ag on Ag ($\theta \approx 10^{-2}, T = 10\ldots60K$) one obtains $E_0 = 65\text{meV}$ and $D_0 = 3.7 \times 10^{-13}\text{cm}^2/\text{s}$. In comparison the evaluation of Chauvineau's result for Au on Au ($\theta \approx 10^{-2}$, $T = 10 \ldots 60K$) leads to $E_0 = 40\text{meV}$ and $D_0 = 4.4 \times 10^{-12}\text{cm}^2/\text{s}$. In the case of gold and silver values of the activation energy E_0 deduced from other methods which ensure that a random walk of monomers is observed (e.g. field ion microscopy) are not available. However, the obtained values of the activation energy are reasonable whereas the pre-exponential factors D_0 are too small compared with values deduced from a simple random walk. For a better agreement a more sophisticated diffusion model is necessary.

Interdiffusion

Beyond the two-dimensional diffusion it is also possible to study special cases
of interdiffusion with conductivity measurements at thin films. In the first
section of this chapter it was shown that the additional resistivity caused by
foreign atoms is unusually higher, if these atoms are bulk impurities instead
of surface impurities. If a metal film covered with a small amount of a foreign
metal is heated up a large irreversible resistivity increase can be observed, at
a characteristic temperature, due to the interdiffusion of the metals (Pariset
1976). This effect can, for instance, be used to study the process of grain
boundary segregation (Munitz, Komem 1980).

The resistivity increase caused by a coverage with silver or benzene differs by about two orders of magnitude. This is the basis of another type of
diffusion experiment: From measurements concerning the surface enhanced
Raman effect the question arises above what temperature silver atoms are
able to diffuse through a condensed benzene film (Grabhorn, Schumacher,
Otto 1990). A six monolayers thick C_6H_6 film is condensed onto a typical
silver film ($d = 20$nm) at 50K. If additional silver ($\theta = 0.02$) is evaporated
on top of the C_6H_6 overlayer film no resistivity variation is observed (Tonscheidt 1990). An isochronal heating follows ($r_A = 0.1$K/s). At a temperature
of about 100K the film resistivity increases significantly, which indicates the
arrival of silver atoms on top of the silver film. Since the induced atomic scale
roughness cannot be annealed until most of the C_6H_6 molecules are desorbed
at 200K, the resistivity remains nearly constant up to this temperature. Above
200K the resistivity decreases as expected until, at 330K, the additional surface defects are mostly eliminated. Therefore the resistivity measurements
can give information about the position of the additionally evaporated silver
atoms in a rather easy way.

5.3 Coverage Dependence

A quantitative description of the coverage dependence of the resistivity should
take into consideration:

- the interaction and the arrangement of the adparticles,

- the multiple scattering approach of the conduction electrons,

- the electronic structure of the metal surface and the adsorbate.

At this time only aspects of these conditions can be mentioned. The following
examples will show that it is a rather complex problem and only the very first
steps have been made:

In the case of gas-coverage experiments a similar behaviour is often observed. The resistivity rises linearly at low coverages and reaches a saturation for high coverages. Here it seems to be possible to transfer evaluation procedures from the scattering of thermal helium atoms to the resistivity measurements (Comsa, Poelsema 1985):

At low coverages every impinging adparticle acts as an additional scattering centre to which an area Σ might be attributed where the conduction electrons are not specularly reflected. Under the assumption that the adparticles impinge statistically and have no mobility the probability that a new adparticle sticks to an already "dereflected" region increases proportionally to the coverage. Consequently the resistivity increase saturates exponentially:

$$\frac{\Delta\varrho}{\Delta\varrho_{\mathrm{max}}} = 1 - \exp(-\Sigma n_{\mathrm{surf}}\theta). \qquad (5.10)$$

The maximum resistivity increase $\Delta\varrho_{\mathrm{max}}$ can be calculated using Sondheimer's approximation (Sondheimer 1952) (see also Sect. 3.2) under the justified assumption that the vacuum side specularity parameter p_{v} decreases from its start value ($\approx 0.8\ldots0.9$) to a lower value. In order to prove the validity of this dependence the values $\Delta\varrho/\Delta\varrho_{\mathrm{max}} - 1$ are drawn versus the O_2-exposure for a 24nm thick silver film in a semi-logarithmic plot (see Fig. 5.8). The data are taken from (Dayal, Finzel, Wißmann 1987). It must be noted that in this experiment the sample temperature is far above the desorption temperature of O_2 on a smooth silver surface. The exposure is adjusted so that an equilibrium between the molecules in the recipient and on the film surface preveils. Therefore the molecules are in a highly excited state, possess a high surface mobility like a two dimensional gas, and are not sticking at special sites. Indeed, the expected linear dependence is obtained in a wide

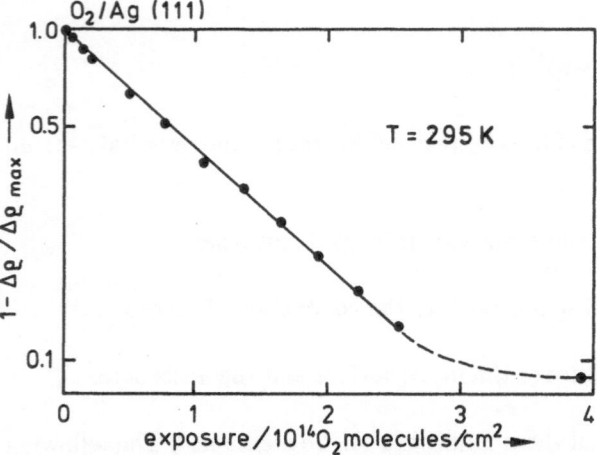

Fig. 5.8. Semi-logarithmic plot of $\Delta\varrho/\Delta\varrho_{max} - 1$ versus O_2-exposure for a silver film ($d = 24$nm), $T = 295$K. Data are taken from (Dayal, Finzel, Wißmann 1987)

range. The deviation at a higher exposure can be explained by the fact that the coverage is not proportional to the pressure in this region. Though this procedure seems to be successful in many cases, this method should not be used here because it is not adequate to describe the scattering of conduction electrons quantitatively.

A multiple scattering approach as described in Sect. 3.2 (Lessie, Crosson 1986) leads to a parabolic dependence (Nordheim-behaviour) of $\Delta\varrho(\theta)$ under the condition that the adparticles successively occupy the points of a two-dimensional surface lattice. Indium atoms evaporated upon an (011)-textured indium film fulfill this condition in the temperature range between 15K and 30K very well (Schumacher, Stark 1984). The measured resistivity variation indeed fits the expected parabolic dependence $\Delta\varrho/\varrho_0 \sim \theta(1-\theta)$ (Fig. 5.9, solid line). The indium base film has a thickness of 20nm. As expected at a coverage

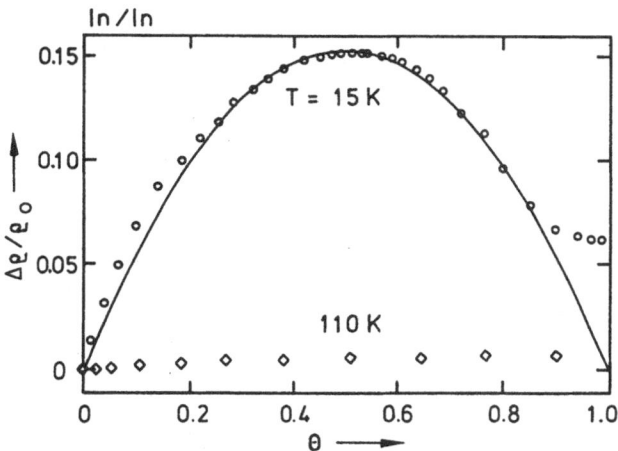

Fig. 5.9. Relative change of the film resistivity $\Delta\varrho/\varrho_0$ due to the evaporation of indium onto an (011)-textured indium film at substrate temperatures of 15K and 110K. Solid line: "Nordheim-behaviour": $\varrho/\varrho_0 \sim \theta(1 - \theta)$

near 1 deviations occur, caused by atoms which are already added in the second adlayer. In the same way as in the case of silver (Fig. 5.1, $T = 350$K) at 110K the indium atoms reach the terrace edges thermally activated and they are incorporated. The resistivity remains nearly constant. The existence of a relative minimum in the resistance of superimposed metal films for a coverage of $\theta \approx 1$ was for the first time shown by Pariset and Chauvineau for several metals evaporated on (111)-textured gold films (Pariset, Chauvineau 1978). Recently, a relative minimum in film resistivity and an approximately parabolic dependence could be obtained for several adsorbed gas species e.g. O_2 on Cu (Dayal, Finzel, Wißmann 1987), C_6H_6 on Ag (Grabhorn et al. 1992), H_2 on Pd (Wedler 1987).

The Influence of Atomic Scale Surface Roughness

Surprising results could be obtained in the case of highly disordered silver films (Holzapfel et al. 1990b). These special films with a thickness of $d = 100$nm were prepared by evaporating the silver on a cold substrate ($T \approx 40$K) without any annealing procedure. Exposures to several different unsaturated hydocarbons ($C_2H_4, C_2H_2, C_6H_6, C_5NH_5$) cause resistance decreases whereas exposures to saturated hydrocarbons (CH_4, C_2H_6) cause resistance increases (Stubenrauch 1987; Holzapfel 1988). In both cases the resistance variation is of the order of 1% and saturates at a coverage far below one monolayer. The result has been explained by an increase or decrease of the electron tunnelling rate at 'tunnelling sites' in porous grain boundaries, if the adsorbates have or do not have, respectively a π^*-orbital (Holzapfel et al. 1990). However, the disadvantage of such cold-condensed films is the unsufficient knowledge of their crystalline structure.

In the following experiment smooth and atomicaly rough silver films are exposed to adspecies (C_2H_6 and C_2H_4) which might affect the local electron density at the surface in a significantly different way (Grabhorn et al. 1992). In the adsorbed state the lowest unoccupied electronic level of C_2H_4 is a π^*-orbital which lies for the adsorbed molecule on a rough silver surface approximately 3.8eV above the Fermi-level E_F of silver [measured by inverse photo emission (Kannen, Reihl, Otto)]. This value is just below the vacuum level. In the case of C_2H_6 the lowest unoccupied molecular orbital is a σ^*-orbital and lies roughly 7.5eV above the Fermi-level E_F and approximately 3.5eV above the vacuum level V [estimated from gas-phase data (Tanaka, Bosten 1985)], see Fig. 5.10 At first, smooth silver films were exposed to 30L

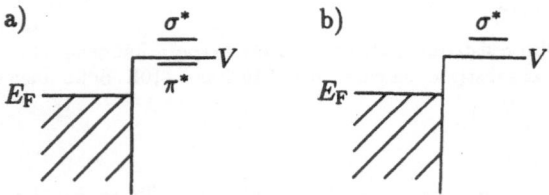

Fig. 5.10. Energy of a single-electron basis function at the silver surface: a) with an unsaturated hydrocarbon adsorbate (e.g. C_2H_4), b) with a saturated hydrocarbon adsorbate (e.g. C_2H_6)

(1L= 1.33×10^{-4}Pa$\times$s) of C_2H_6 and C_2H_4 at a substrate temperature of 50K. The resulting resistivity increases are plotted in Fig. 5.11 and 5.12). In the case of C_2H_6 a weak minimum at about 4L indicates the completion of a first ordered adsorbate layer. The admolecules cover the whole surface. (The comparably high value of 4L is caused by a mask in front of the film which limits the solid angle of the molecules striking the surface.)

In a second run the surfaces of new films are disturbed by approximately half a monolayer of silver atoms at 50K onto the smooth films. At such a low temperature the adatoms are monomers or form small islands on the terraces. During the deposition of silver the film resistivity increases by about $\Delta\varrho = 0.8\mu\Omega$cm. Afterwards the films are exposed to C_2H_6 and C_2H_4 in an equivalent way. The changes of the resistivity obtained now are also plotted in Fig. 5.11 and 5.12. For C_2H_6 a higher resistivity increase is obtained with a pronounced relative minimum at a coverage significantly below one monolayer, whereas in the case of C_2H_4 a fraction of admolecules less than a monolayer causes a sharp and well reproducible resistivity decrease followed by an increase similar to the 'smooth case'. Obviously the leading decrease or increase of resistivity is caused by a small fraction of molecules which cover sites in contact to the silver adatoms or islands due to the higher binding energy at those sites. At

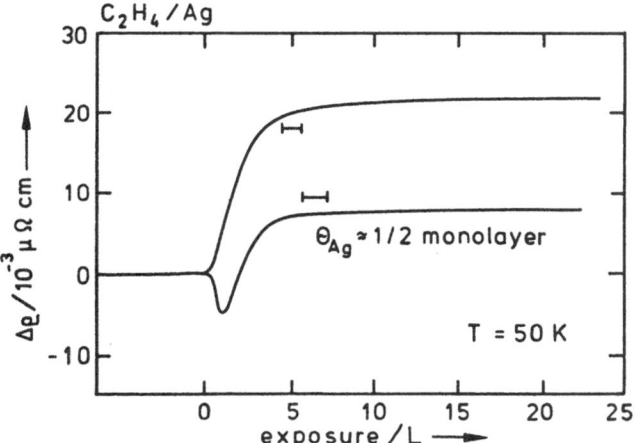

Fig. 5.11. Resistivity changes $\Delta\varrho$ of smooth and atomically rough silver films ($d = 30$nm) due to an exposure of C_2H_4 (1L= 1.33×10^{-4}Pa×s) (Grabhorn et al. 1992)

a smooth film surface both gas-species act as scattering centres. However, at an atomically rough film a first small amount of admolecules shows a specific effect. Distortions of the local electron density caused by silver adatoms can be reduced by molecules with an unoccupied level below the vacuum level. Otherwise the molecules enhance the distortion. A detailed interpretation is just under consideration and will be given elsewhere (Grabhorn et al. 1992). These types of experiments seem to be suitable to shed more light upon the influence of adsorbed molecules on the resistivity of a thin metal film, though theoretical calculations and more measurements are necessary.

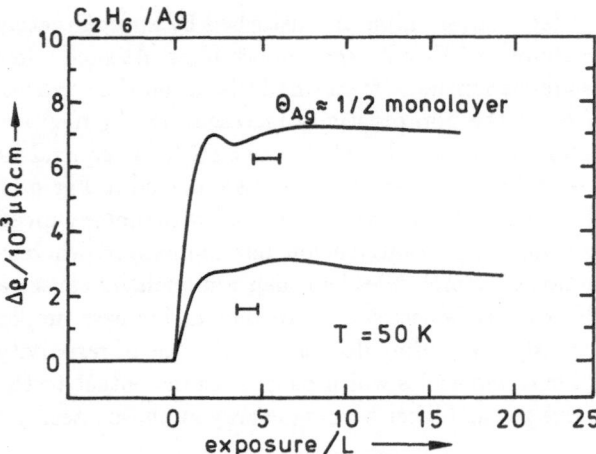

Fig. 5.12. Resistivity changes $\Delta\varrho$ of smooth and atomically rough silver films ($d = 30$nm) due to an exposure of C_2H_6 (1L= 1.33×10^{-4}Pa×s) (Grabhorn et al. 1992)

5.4 Film Growth

Epitaxial systems belong to the main subjects of material science due to their application as optical coatings, as passive layers and as components in semi-conductor devices. The rising quality requirements call for an in-situ control of the growth process. For example, in order to obtain a surface with a min-imum of defects, it is necessary to stop the evaporation process just when a monolayer is completed. The most successful techniques suitable for an in-situ control of the epitaxial growth are the reflected high energy electron diffrac-tion (RHEED) and the scattering of thermal helium atoms (TEAS). Short introductions will be given in order to compare the results of the conductivity measurements with these techniques:

Koziol, Lilienkamp and Bauer (Koziol, Lilienkamp, Bauer 1987) were the first to report successful in-situ RHEED studies of the epitaxial growth of met-als on metallic substrates. They have scattered a beam of electrons (15keV) at the surface of the growing film and have measured the current of the spec-ularly reflected beam as a function of the deposition time. Figure 5.13 shows the results for the epitaxial system Ni on W(110). The oscillations of the cur-rent can be attributed to the completions of monolayers during the epitaxial growth. The abrupt changes of the amplitude at the beginning are caused by the lattice mismatch between Ni and W. It is evident that this method can be used to monitor the growth process. However, a quantitative description of the RHEED-intensities requires a multiple scattering approach which com-plicates an evaluation. In comparison, the scattering of thermal helium atoms can be explained with the help of a simple kinematic approach. In contrast the experimental expense is much higher. In order to monitor the epitaxial growth the specularly reflected beam is detected. Figure 5.14 is taken from a

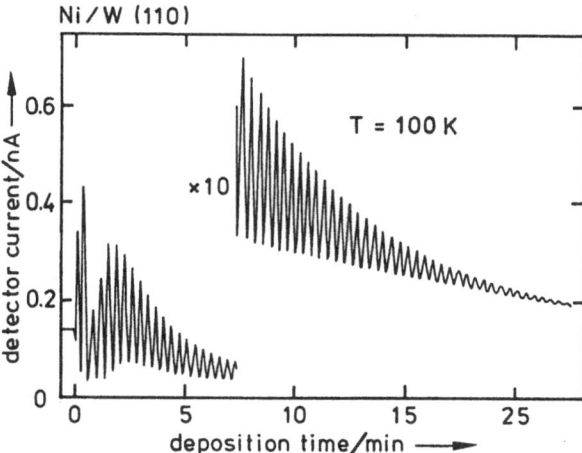

Fig. 5.13. Intensity of the specularly reflected beam of high energy electrons ($E_{kin} = 15 keV$) as a function of the deposition time for the epitaxial system Ni on W(110) (Koziol, Lilienkamp, Bauer 1987)

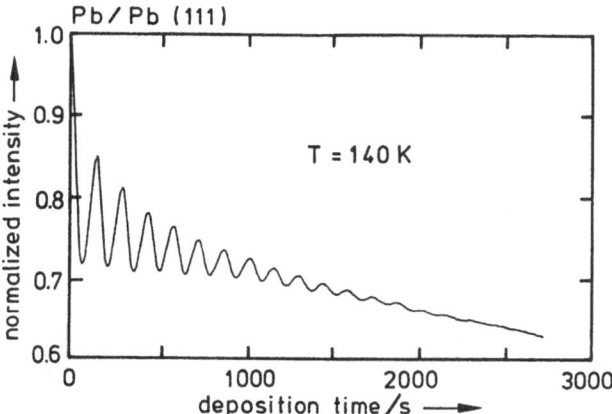

Fig. 5.14. Normalized intensity of the specularly reflected beam of thermal helium atoms as a function of deposition time for the epitaxial system Pb on Pb(111) (Hinch et al. 1989)

work of Hinch, Koziol, Toennies and Zhang (Hinch et al. 1989). As an example it presents the results obtained from the epitaxial system Pb on Pb(111). The oscillations can also be attributed to the layer by layer growth. The fact that TEAS can be explained within a kinematic approach and that it is a true surface scattering method leads to a rather simple and direct interpretation concerning the origin of the oscillations.

The preparation of the single crystal is done in general with the aim of producing a rather smooth surface. The result is a surface which consists mainly of two different consecutive atomic-planes. The extended terraces are nearly perfect and the most common type of surface defects are the remaining terrace edges. Here the terrace-steps go up and down alternativly. However, in the case of the thin films described above the terrace-steps form a 'staircase' on top of each crystallite. For the scattering prossess, it must be distinguished between two conditions [see for example (Miguel et al. 1988)].

Out-of-phase condition means that scattering waves originating from both atomic-planes lead to a destructive superposition at the detector. The signal reaches a minimum if each plane forms 50% of the surface. During epitaxial growth the ratio of the uppermost to the next atomic plane oscillates, which causes oscillations of the specular TEAS-intensity. Under these condition in first order the signal is not influenced by the density of edges and the amplitude of these oscillations increases with increasing substrate temperature. Substrates (e.g. the films used here) which pocess a sequence of terraces like a staircase are unsuitable to observe these oscillations.

Under the *in-phase-condition* the scattered waves from the (two) uppermost atomic-planes interfere constructively at the detector position. A maximum in the intensity is correlated with a minimum of surface defects, e.g. terrace edges. In a certain temperature range the admaterial grows in two dimensional aggregates and the edge density increases until it reaches a maximum at $\theta \approx 0.5$. With rising coverage the islands coalesce and the defect density decreases until $\theta \approx 1$. This process is repeated layer by layer. The curve in Fig. 5.14 is of this type (Zhang 1991). In principle it should be possible to obtain this type of TEAS-oscillations using the films described here as substrates. The oscillations observed with RHEED can only be obtained under the in-phase-condition (Ferron et al. 1989). However, they are not so easy to explain, because a multiple scattering approach is necessary.

Oscillations of the Film Resistivity Caused by a Layer by Layer Growth

Oscillations of the film resistivity during the film growth comparable to the last type can be observed in an increasing number of epitaxial systems. Two examples are given in Fig. 5.15 and 5.16. Here the relative change of the resistivity is plotted versus the thickness d_2 of a covering film. In a certain temperature range (In: $T \approx 15K$ and Bi: $T \approx 78K$) the film resistivity oscillates during the evaporation of additional metal onto the smooth base film. An evaluation along the line indicated in Sect. 3.3 shows that in the case of indium the resistivities ϱ_1 and ϱ_2 of the base film and the covering film are nearly equal, independent of the preparation temperature (Schumacher, Stark 1984). Therefore no significant differences in the defect density between the the parts of the film can be recognized.

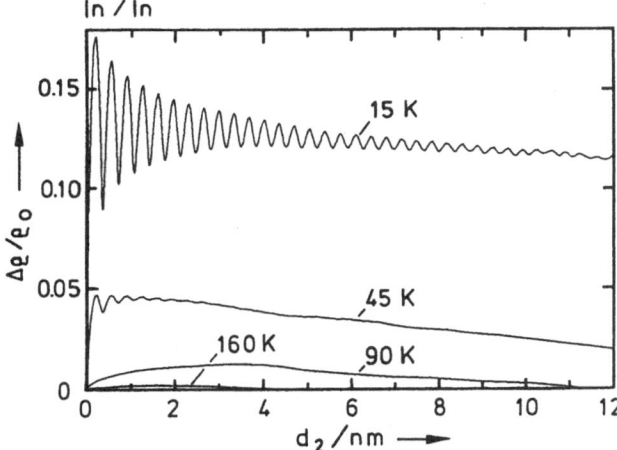

Fig. 5.15. Relative resistivity change $\Delta\varrho/\varrho_0$ versus overlayer thickness d_2 at different substrate temperatures for the epitaxial system In on In. Thickness of the base film: $d_1 = 20$nm

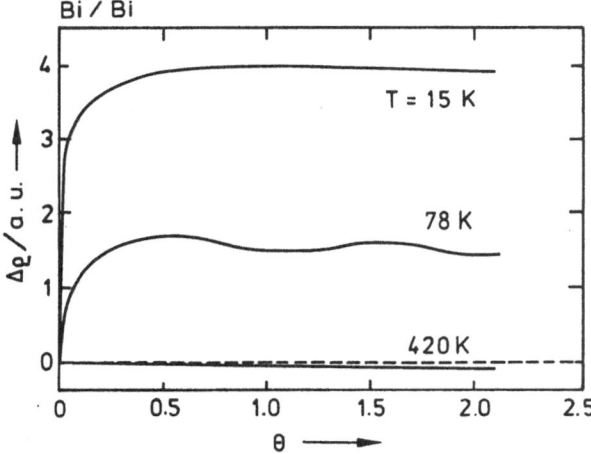

Fig. 5.16. Change of the resistivity $\Delta\varrho$ in arbitrary units versus coverage θ at different temperatures for the epitaxial system Bi on Bi. Thickness of the base film: $d_1 = 25$nm (Heimlich 1983)

Computer Simulations of the Growth Process

In order to elaborate a more detailed picture of the growth process computer simulations can be useful. In the computer simulations the surface of an individual crystal of the film is represented as a field of 128 by 128 integer elements which represent the possible atom positions in a simple cubic surface lattice. It becomes apparent that no size dependence can be observed. Nevertheless, to avoid boundary effects periodic boundary conditions are implemented. In order to incorporate the varying surface conditions on different crystallites, the results are averaged over 20 computer runs with different pre-

coverages of the initial surface. The film growth consists of two concurrent
processes: Condensation and the random walk of the adparticles. The reevap-
oration of the atoms can be neglected. Both processes are controlled by a
random generator. Only monomers are allowed to jump. As soon as an adparticle has reached a site where at least one of its next neighbour positions is
occupied its random walk stops here. This means that a dimer is a stable nu-

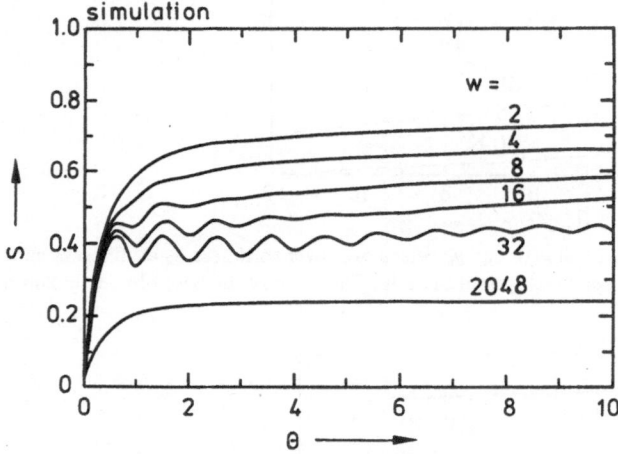

Fig. 5.17. Results of the computer simulations (density of atomic steps S as a function of
the coverage θ). The quantity w is connected with the mobility of the monomers (see text)

cleus (Zinsmeister 1966). This behaviour represents the energetic conditions
for a metal atom on a clean metal surface in a simplified but sensible way.
The simulation delivers the density of atomic steps S on the artificial surface
as a function of the coverage θ. The quantity S behaves in the same way
as the type of surface roughness does which cause diffuse electron scattering
(Schumacher, Stark 1984). Figure 5.17 shows the result of these calculations.
The parameter w represents the mobility of the monomers on the surface.
It is defined as the number of permitted jumps of the monomers per impinging particle. In addition Fig. 5.18 images an area of the artificial surface
in the surroundings of an original terrace ledge. The numbers represent the
number of atoms which stick above a plane arbitrarily set to zero. Part a)
of Fig. 5.18 shows the initial stage of the process, two terraces separated by
a one-atomic step, the other parts b) - d) belong to a coverage $\theta = 1$. At
sufficiently low temperatures the impinging adatoms can only make about
2 jumps ($w = 2$) due to the released binding energy (McCarroll, Ehrlich
1963). The "surface roughness" S increases steeply and reaches a maximum
for about 1 to 2 monolayers. In the experiments this situation is realized for
Bi on Bi at 15K. Part b) of Fig. 5.18 gives an image of the surface structure
at a coverage of $\theta = 1$. The original surface structure is completely covered by
atomic-scale roughness. This explains the significant resistivity increase. At a
certain mobility ($w = 32$) the density of atomic steps S of the artificial surface

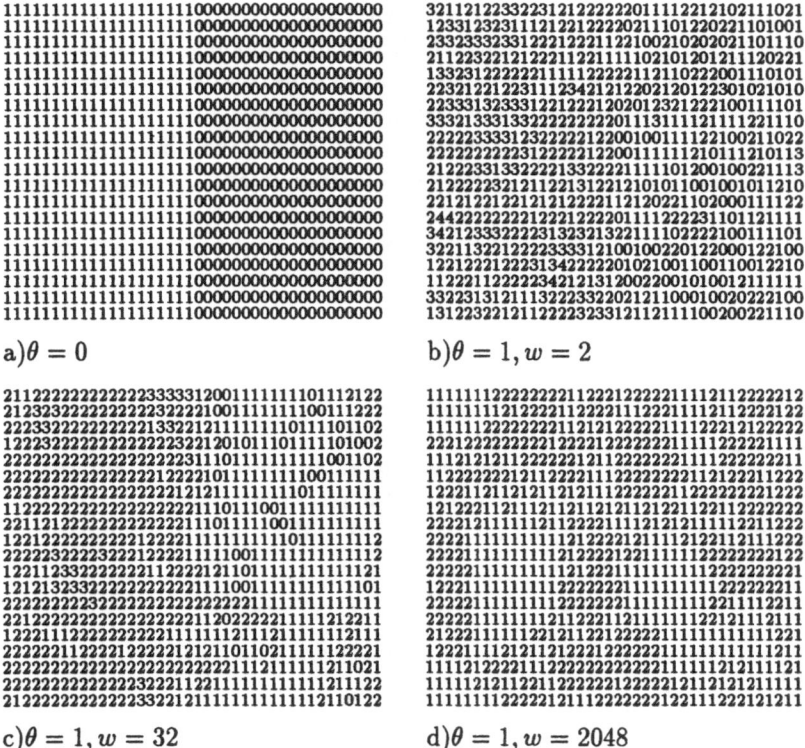

Fig. 5.18. Area of the artificial surface at $\theta = 0$ and $\theta = 1$ for different mobilities w. Start condition: two terraces with a one-atomic step

oscillates during the film growth comparable to the resistivity behaviour (Bi: $T \approx 77\mathrm{K}$ and In: $T \approx 15\mathrm{K}$). With the help of Fig. 5.18 it can be explained what happens. Apart from "point defects" the original surface structure has been reproduced, while the film has grown one monolayer. The terrace edge can still be recognized and the terraces exhibit large smooth areas. In this case the simulation curve also fits the measured resistivity curves well. This proves that the conception formulated with the computer simulation indeed approaches the true processes on the film surface. With higher mobility the probability that an impinging atom reaches the terrace edge increases. So the local surface structure influences the growth process. If the adatoms are not able to move along the terrace edges but have a high mobility ($w = 2048$) on the terraces, the mean roughness S increases slightly because the terraces are replaced by two dimensional islands (Fig. 5.18 d). There is no longer oscillatory behaviour. A corresponding experimental curve is given in Fig. 5.15 (In on In(011), $T = 90\mathrm{K}$). It is evident that if the admaterial is slowly evaporated at the annealing temperature of the base film, the surface structure which is an equilibrium state at this temperature cannot be changed. Therefore constant or slightly decreasing resistivity is observed (In: $T \approx 160K$, Bi: $T \approx 420K$). Computer simulations which permit a mobility of atoms along the edges show that the terrace edges seem to walk across the surface, but the surface structure and the value S remain unchanged.

Corresponding oscillations of the resistivity during the film growth have been observed for tin, lead and gallium besides indium and bismuth, whereby in most cases noble-metal films can also serve as substrates. The discussion has shown that the oscillations observed in the resistivity during film growth are closely related to the oscillations of the reflected beam in TEAS obtained under the in-phase condition.

Annealing of Atomic Scale Roughness

In an additional experiment indium is evaporated onto a (011)-textured indium film at 15K, but at the maximum of the resistivity change ($\theta = 0.5$) the evaporation is stopped and the film is held at 15K for about 30min. During

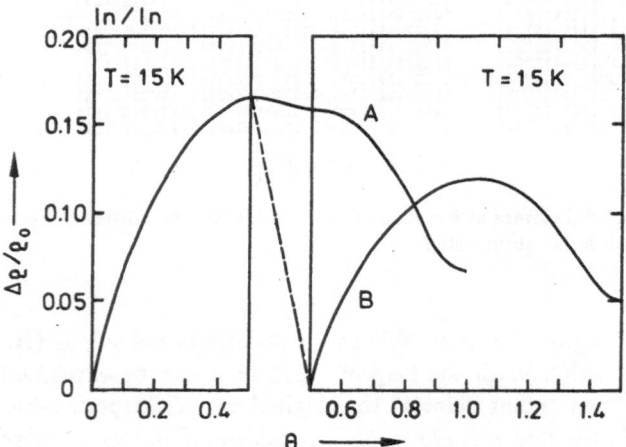

Fig. 5.19. Resistivity variation of a (011)-textured indium film ($d_1 = 20nm$) during evaporation of additional indium at 15K. A) The evaporation is interrupted for about 30min at $\theta = 0.5$ and the film is held at 15K. B) During the interruption the film is heated up to 160K and cooled down again to 15K

this time relaxation processes in the overlayer cause a small resistivity decrease (see Fig. 5.19). If the evaporation is continued (A) the film resistivity decreases until a minimum is reached at $\theta = 1$ as it is obtained without a stop. In a second experiment the film is heated up to the annealing temperature of the base film ($T_A = 160K$) during the break. The additional resistivity caused by the evaporation of the first half of a monolayer disappears completely, which can be seen when the film is cooled down to 15K. Now the original state of the film surface has been reached again and the following evaporation starts with a rising atomic scale roughness (B). These experiments corroborate the given explanation and show how the growth process can be monitored by the resistivity measurements.

Lattice Mismatch and Superstructures

Chauvineau and Maliére (Maliére 1986; Chauvineau, Maliére 1985) investigated the thermal behaviour of indium adlayers on gold films. With a very precise quartz crystal thickness monitor he has also determined the exact position of the first minimum for this system. X-ray diffraction studies show that indium grows along the [101]-direction on the (111)-textured gold films unlike the previous case (Pariset 1976). Maliére shows (see Fig. 5.20) that,

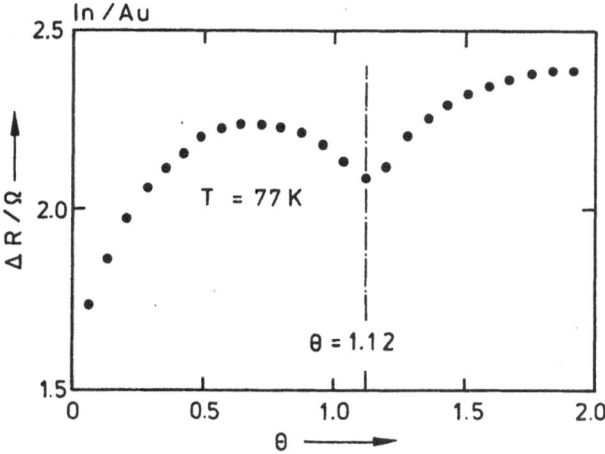

Fig. 5.20. Resistance variation $\Delta \varrho$ of an (111)-textured gold film during evaporation of indium (Marliére 1986)

due to the lattice misfit between gold and indium, about 12% more indium atoms are necessary to form the first indium layer on the gold film compared with the number of indium atoms forming a regular In(101)-plane. This value lies halfway between the number of gold atoms per area in a (111)-plane ($1.39 \times 10^{15} \text{cm}^{-2}$) and the density of indium atoms in a (101)-plane ($1.04 \times 10^{15} \text{cm}^{-2}$). Obviously the gold substrate forces the indium atoms into a more compact interface plane. It is well known that at low coverages above room temperature lead forms a $\sqrt{3} \times \sqrt{3}$ R30° superstructure on a Au(111) surface (Perdereau, Biberian, Read 1974). Pariset and Chauvineau have observed the appearance of this superstructure by measuring the resistance of a (111)-textured gold film during coverage with lead. At a temperature of 423K they have obtained a well recognizable saddle point at $\theta = 1/3$ as an indication of an ordered adstructure. For a lower temperature $T = 293$K a plateau exists which starts at $\theta = 1/3$ and is limited by a minimum at $\theta = 1$.

This example shows that the combination of LEED and dc-resistivity studies can be a fruitful subject in future. Textured films are less suitable in this case, monocrystalline base films are necessary. The high reproducibility of the layer by layer growth experiments permits their use as an indicator of changes of the substrate film surface, for example by a precoverage.

The Influence of a Precoverage

What happens if the indium base film is already covered with foreign atoms before the evaporation of an indium overlayer film starts? Argon, oxygen and silver are used as precoverage. The exposure or evaporation takes place at a sample temperature of 15K and the deposited amount corresponds, in all cases, to a coverage of approximately half the surface area. The results concerning the three adspecies diverge significantly and a qualitative explanation can easily be given. The precoverage of argon causes a decrease of the amplitudes of the layer-by-layer growth of about 30%. A measurable 'phase shift' of the oscillations is not observed. Predeposited silver atoms are incorporated as expected, and the resistivity oscillations are shifted analogously to Fig. 5.19 (curve B). A different result is obtained in the case of an oxygen preadsorption. The evaporation of the first half of a monolayer of indium leads to a resistivity increase of 30% which is much more than in the clean case. The mass of one more monolayer must be evaporated before the first weak minimum occurs. The following oscillations are weak but their period length corresponds to $\Delta\theta = 1$.

Superimposed Noble-metal Films

Figure 5.21 shows the normalized conductance of superimposed silver films as a function of the overlayer thickness d_2. The base film thickness d_1 is always 20nm. The temperature during coverage has been varied between 10K and the annealing temperature of the base film (350K). Here a special plot R_0/R versus d_2 is presented in order to explain a useful evaluation procedure. Under the assumption that the resistivities ϱ_1 and ϱ_2 of the base film and the covering film respectively are constant and the layer system can be treated as a parallel connection, R_0/R is a linear function of d_2 . Using (3.16) the ratios ϱ_1/ϱ_0 and ϱ_2/ϱ_0 can be determined by this linearity. Here such an evaluation is possible for a covering film thickness above $d_2 \approx 4$nm (see Fig. 5.21). Some results are compiled in Table 5.1. If the admaterial is condensed at 350K the resistivities of both parts of the film equal the resistivity of the uncovered base film ϱ_0 . At 10K the resistivity of the base film increases by $\approx$40% caused by the coverage with a highly disordered film of high resistivity. Both cases have just been discussed with the help of computer simulations. However, in the intermediate temperature range for silver on silver as well as for the other

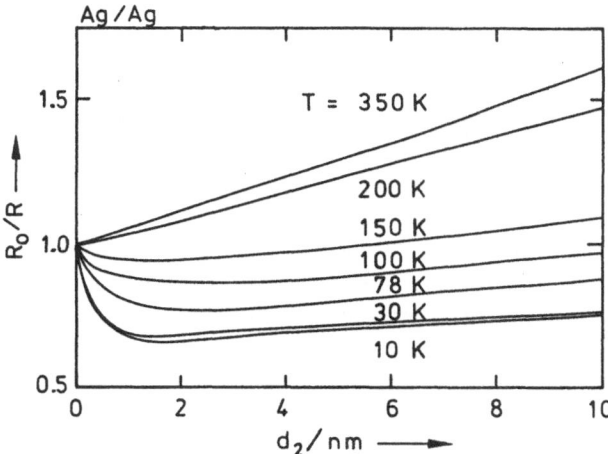

Fig. 5.21. Normalized conductance R_0/R versus covering film thickness d_2 for the deposition of silver on (111)-textured silver films

Table 5.1. Resistivity ratios of the parts of superimposed silver films for different deposition temperatures of the covering films (evoluation: see text)

T/K	ϱ_1/ϱ_0	ϱ_2/ϱ_0
10	1.4±0.1	5.3±0.1
78	1.3	2.5
350	1.0	0.9

noble-metals (Pariset 1976; Fischer 1980; Fischer, Minnigerode 1981; Schumacher, Stark 1982; Schumacher 1983; Fischer, Moske, Minnigerode 1983) no oscillations, due to a varying atomic scale roughness, appear. In all cases a smooth but significant resistivity increase (decrease in conductance R_0/R) can be observed with a maximum at a coverage between 1 and 2nm. For gold high resolution TEM studies give an explanation which can be generalized to copper and silver. In the interesting temperature regime an atom impinging on top of a small two-dimensional island exhibits the tendency to remain there. This is no perfect island-growth [for a detailed discussion see (Meinel, Klaua, Bethge 1988)]. However, atoms are added in the second layer far before the first layer is completed. This behaviour destroys a synchronous layer by layer growth and the maximum of surface roughness will not be reached until several layers are added.

Metallic Superlattices

Metallic superlattices have gained of importance during the last years. They are especially of interest as Bragg-diffractors for soft X-rays (Spilles, Segmüller 1989) or as samples with special electric properties (Hoffmann, Kücher 1987). The aim is to show that the resistivity measurements can be used to control the growth of a metallic superlattice in-situ. The system shown here as an example starts with a base film of approximately 85 monolayers indium whose thickness is measured by a conventional quartz crystal thickness monitor. This film is prepared and annealed as described above. Afterwards, alternating layers of tin and indium are evaporated at a substrate temperature of 20K, controlled by the dc-resistivity measurement as shown in Fig. 5.22. This multilayer is finished with ten monolayers of indium. The heights of the

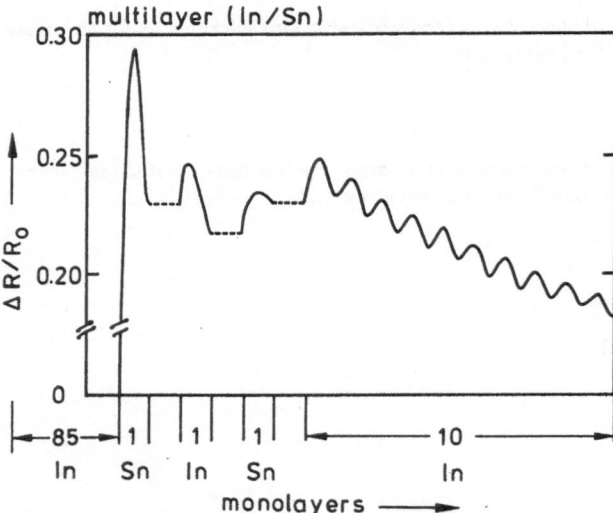

Fig. 5.22. Epitaxial growth of an In/Sn-superlattice monitored by a dc-resistivity measurement

resistivity changes prove the enormous sensitivity and the simplicity of this technique, which is open to many applications. It can be expected that such superlattices will exhibit specific electric and electronic features.

Quantum Oscillations in Metallic Hetero-films

In some cases the resistance variation of a metal film during the overlayer growth shows an additional pseudo-periodic structure (Chauvineau, Pariset 1976; Pariset, Chauvineau 1978; Schumacher, Stark 1982; Schumacher, Stark 1983; Schumacher 1983). Pariset and Chauvineau (Pariset, Chauvineau 1978) were the first who mentioned these structures in the case of indium on gold.

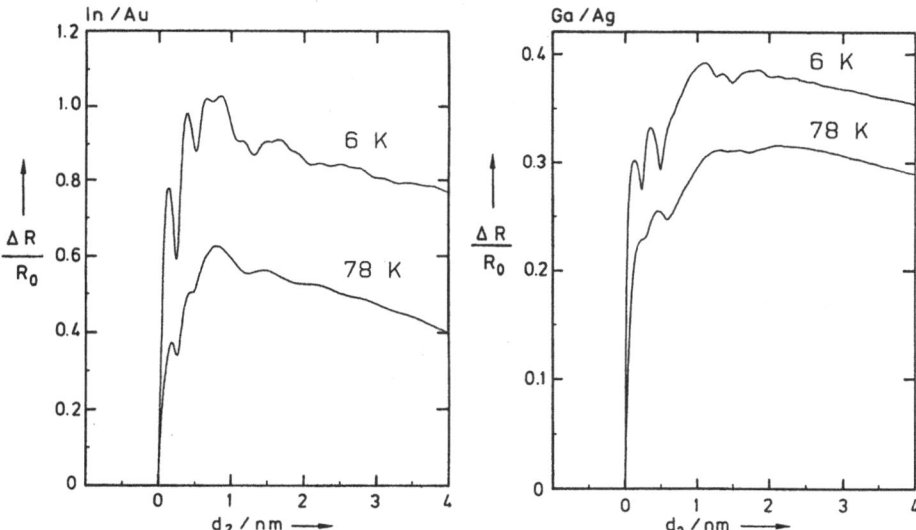

Fig. 5.23. Relative change in resistance versus overlayer thickness measured at the layer system In on Au and Ga on Ag ($T = 6$K and 78K)

They explained them as caused by the quantum-size-effect in the heterogeneous superimposed films. Figure 5.23 gives two examples of different heterofilms showing these structures. The structures only appear in the case of heterogeneous superimposed films. An epitaxial growth of the overlayer film is a necessary precondition. In a certain temperature range the amplitude increases with lower temperature. The periodic length depends on the overlayer film and not on the base film material. The oscillations are caused by the appearence of standing electron waves in the covering film. The phenomenon is very similar to the quantum-size-effect in single metallic films as described for instance by Sandomirskii or Schulte (Sandomirskii 1967; Schulte 1976; Schulte 1977). In the case of a single metallic film the resistivity is a pseudo-periodic function of the film thickness with a period length of approximately $\lambda_F/2$ (Trivedi, Ashcroft 1988). Since for normal metals $\lambda_F/2$ is of the order of the lattice constant (see Sect. 4.2) the quantum-size-effect should influence the resistivity of a continuous metal film in the thickness range between 0 and $\approx$3nm at low temperatures. However, a metallic film of this thickness prepared on a dielectric substrate has nearly always a discontinuous structure. Therefore the resistivity is dominated by the tunnelling process from island to island. Only a few experimentalists have tried to overcome this problem. Hoffmann and Fischer (Hoffmann, Fischer 1976; Fischer, Hoffmann 1980) have developed a sophisticated measuring circuit to register the derivative $\partial\varrho/\partial d$ as a function of the film thickness d in order to surpress the large continuous thickness dependence of the resistivity. Jalochowski and

Bauer (Jalochowski, Bauer 1988) have used an especially prepared silicon substrate $(Si(111)/Au(6\times6))$ to prevent the formation of islands. Nevertheless, the observed structures are less pronounced. The advantage of the experiment described here is the possibility to form very thin continuous metal films on a metallic substrate or substrate-film.

It is useful to compare these measurements with an experimental approach of Jonker et al. (Jonker, Bartelt, Park 1983; Jonker, Park 1984). They have measured the transmission coefficient of very low energy electrons $(E_{kin} \leq 10eV)$ normally incident on epitaxial Cu(111) and Ag(111) films on a W(110) substrate. The transmitted current is modulated by interference between the electron wave scattered from the vacuum-metal and the film-substrate boundary. In the experiment the film thickness is fixed $(1\ldots 7.5nm)$ and the electron wavelength is changed by tuning the kinetic energy. In the case of the resistivity measurement the wavelength of the electrons responsible for the conduction process is fixed $\lambda \approx \lambda_F$. However, the film thickness varies during the growth of the overlayer. In both types of experiments it is necessary that the boundaries are sufficiently smooth and less extended. Additionally, a certain reflectivity of the metal-metal interface must be assumed. A simple potential step as claimed by Jonker and Park (Jonker, Park 1984) does not contain all sources of reflection. For wave propagation it is necessary that the wave functions match at the interface. If the lattice periodicities of both metals are quite different the wave is at least partially reflected even in the absence of a significant potential step. Especially, in the case of gold and indium metallic compounds can exist at the interface (Kepper et al. 1986) and might cause localized charges. This effect can be modelled by an additional potential wall.

The film thickness does not grow continuously but in steps of atomic layers. Therefore an interrelation between the growth process and the quantum interference phenomenon can be supposed. Feibelman (Feibelman 1983; Feibelman, Hamann 1984) has carried out self-consistent calculations for small slabs with up to seven layers. He calculated the surface energies and found that the value of the two layer slabs is less than the value of one, three, four or six layer slabs. This result supports the presumption that the quantum-size-effect might influence the growth mechanism. In a simple picture a number of layers whose thickness is comensurate with $\lambda_F/2$ is favoured, whereas an incomensurate thickness is avoided. One will expect that the amplitude of the oscillations caused by the layer by layer growth is no more monotonously damped as it is in the case of In on In (resistivity measurement) or Pb on Pb(111) (TEAS). For TEAS from Pb on Cu(111) Hintch et al. have observed for the first time that the amplitude 'beats' and that the curve looks like a sequence of 'W'- and 'M'-like structures (Hinch et al. 1989). A similar behaviour can be seen in Fig. 5.23 for Ga on Ag.

However, additional experiments and a more elaborated theoretical basis are necessary to understand this behaviour. Possible applications of the quantum-size-effect in metal films have been discussed by Elinson et at. (Elinson et al. 1972).

6. Final Remarks

If the thickness of a thin continuous metal film is of the same order of magnitude as the mean free path defined in the Drude-Sommerfeld Model of metallic conductivity, scattering events of the conduction electrons at the surface can contribute significantly to the resistivity. This fact can be used to study surface processes with the help of comparatively simple dc-resistivity measurements. Since the resistivity of a thin metal film can be influenced by many different phenomena, emphasis must be laid on a careful preparation and a complete characterization of the thin films. Especially, in order to do surface physics it is neccessary to prepare the films under ultra high vacuum conditions and to use films with well defined surfaces. The method is based on the justified assumption that a defect free closed packed surface acts like a 'mirror' for conduction electrons. Minute amounts of adparticles can disturb this condition and lead to a significant resistivity increase. The dc-resistivity measurements have turned out to be a useful additional tool of surface physics to investigate processes like condensation, adsorption, desorption, surface diffusion, formation of compressed layers and superstuctures, catalytic reactions and first stages of crystalline growth. In several cases the epitaxial growth of metallic films and multilayers can be monitored similarly to RHEED (reflected high energy electron diffraction) and TEAS (thermal energy atom scattering), but with much less technical expense. The relation between the electronic structure of the adparticle and its specific influence on the film resistivity is still under consideration. Here, new experimental approaches and recently developed theoretical images will lead to more progress in the near future.

References

Abrahams, E., P.W. Anderson, D.C. Licciardello and T.V. Ramakrishnan, 1979: Phys. Rev. Lett. **42**, 673. Sect. 2.2

Aharonov, Y. and D. Bohm, 1959: Phys. Rev. **115**, 485. Sect. 2.1

Argile, A. and G.E. Rhead, 1989: Surface Sci. Reports **10**, 277. Sect. 4.1

Arlinghaus, F.J., J.G. Gay and J.R. Smith, 1980: Phys. Rev. B **21**, 2050. Sect. 3.1

Arlinghaus, F.J., J.G. Gay and J.R. Smith, 1981: Phys. Rev. B **23**, 5152. Sect. 3.1

Ashcroft, N.W., N.D. Mermin, 1981: Solid State Physics (Holt-Saunders International Editions, Tokyo). Sect. 2.1

Belzak, V., M. Kedro and A. Pevala, 1974: Thin Solid Films **23**, 305. Sect. 3.3

Bergmann, G., 1984: Physics Reports **107**, 1. Sect. 2.2

Besocke, K., 1987: Surface Sci. **181**, 145. Sect. 4.2.3

Boato, G., P. Cantini and R. Tatarek, 1976: J. Phys. F **6**, L237. Sect. 3.2

Brückner, M., 1982: Examensarbeit, University of Düsseldorf F.R.G. Sect. 4.1

Bülow, H. and W. Buckel, 1956: Z. Physik **145**, 141. Sect. 4.1

Burton, W.K., N. Cabrera, F.C. Frank, 1951: Phil. Trans. Roy. Soc. **243**, 299. Sect. 5.2

Chauvineau, J.P., 1980: Surface Sci. **93**, 471. Sects. 2.4, 5.2

Chauvineau, J.P. and C. Marliére, 1985: Thin Solid Films **125**, 25. Sect. 5.4

Chauvineau, J.P. and C. Pariset, 1973: Surface Sci. **36**, 155. Sect. 2.4

Chauvineau, J.P. and C. Pariset, 1976: J. de Physique, **37**, 1325. Sects. 2.4, 5.4

Chopra, K.L., 1969: *Thin Film Phenomena*, (McGraw - Hill Inc., New York). Sect 2.4

Chopra, K.L. and M.R. Randlett, 1967: J. Appl.Phys. **38**, 3144. Sect. 2.4

Comsa, G. and B. Poelsema, 1985: Appl.Phys. A **38**, 153. Sects. 3.2, 5.3

Crank, J., 1967: *The Mathematics of Diffusion*, (Oxford University Press, London). Sect. 5.2

Croce, P., G. Devant and M.F. Verhaege, 1965: *Basic Problems in Thin Film Physics*, Proc. Intern. Symp. Clausthal–Göttingen Ed.: R. Niedermayer and H. Mayer (Vandenhoek and Ruprecht Göttingen) p.194. Sect. 4.2.2

Dayal, D., H.U. Finzel and P. Wißmann, 1987: *Resitivity Measurements on Pure and Gas Covered Silver Films*, in Thin Metal Films and Gas Chemisorption, Studies in Surface Science Catalysis, Vol. 32, Ed.: P. Wißmann (Elsevier, Amsterdam, Oxford, New York, Tokyo). Sects. 2.4, 3.2, 5.1, 5.3

DESAG, 1989: Dr. Beer, Firma DESAG (Deutsche Spezialglas AG), private communication. Sect. 4.1

Doezema, R. and J.F. Koch, 1972: Phys. Rev. B **5**, 3866; Phys. Rev. B **6**, 2071. Sect. 3.1

Dimmich, R., 1988: Thin Solid Films **158**, 13. Sect. 3.3

Dimmich, R. and F. Warkusz, 1983: Thin Solid Films **109**, 103. Sect. 3.3

Dimmich, R. and F. Warkusz, 1986: Thin Solid Films **135**, 43. Sect. 3.3

Ehrlich, G. and K. Stolt, 1980: Ann. Rev. Phys. Chem. **31**, 603. Sect. 5.2

Elinson, M.I., V.A. Volkov, V.N. Lutikii and T.N. Pinsker, 1972:Thin Solid Films **12**, 383. Sect. 5.4

Elsom, K.C. and J.R. Sambles, 1981: J. Phys. F **11**, 647. Sect. 3.2

Euceda, A., D.M. Bylander and L. Kleinman, 1983: Phys. Rev. B **28**, 528. Sect. 3.1

Feibelman, P.J., 1983: Phys. Rev. B **27**, 1991. Sects. 2.2, 5.4

Feibelman, P.J. and D.R. Hamann, 1984: Phys. Rev. B **29**, 6463. Sects. 2.2, 5.4

Ferron, J., J.M. Gallego, A. Cebollada, J.J. De Miguel and S. Ferrer, 1989: Surface Sci. **211**, 797. Sect. 5.4

Fischer, B., 1980: Thesis, University of Göttingen F.R.G. Sects. 3.3, 5.1

Fischer, B. and G. v. Minnigerode, 1981: Z. Physik B **42**, 349. Sect. 5.4

Fischer, B., M. Moske and G. v. Minnigerode, 1983: Z. Physik B **51**, 327. Sect. 5.4

Fischer, G. and H. Hoffmann, 1980: Z. Phys. B **39**, 287. Sect. 5.4

Fischer, W. and P. Wißmann, 1982: Appl. Surface Sci. **11**, 109. Sect. 4.2.2

Fuchs, K., 1938: Proc. Cambridge Phil. Soc. **34**, 100. Sects. 2.4, 3.2

Garcia, N. and F. Flores, 1984: Physica B **127**, 137. Sect. 3.1

Garcia, N., Y.H. Kao and M. Strongin, 1972: Phys Rev. B **5**, 2029. Sect. 2.2

Gerlach, E., 1984: Phys. Stat. Sol. (b) **121**, 757. Sect. 3.2

Gerlach E., 1990: private communication. Sect. 3.2

Gerlach, E. and P. Grosse, 1977: Festkörperprobleme **17**, 157. Sect. 3.2

Geus, J.W., 1971: *The Influence of Adsorption on Electrical and Magnetic Properties of Thin Metal Films* in: Chemisorption and Reactions on Metallic Films I, Ed.: J.R. Anderson (Academic Press London, New York). Sect. 5.1

Glocker, R., 1958: Materialprüfung mit Röntgenstrahlen (Springer, Berlin). Sect. 4.2.2

Grabhorn, H., D. Schumacher and A. Otto, 1990: Verh. DPG, O **22.1**, 1149. Sect. 5.2

Grabhorn, H., A. Otto D. Schumacher and B.N.J. Persson, 1992: Surface Sci. in press. Sect. 5.3

Greene, R.F., 1966: Phys. Rev. **141**, 687. Sect. 3.2

Greene, R.F. and J. Malamas, 1973: Phys. Rev. B **7**, 1384. Sect. 3.2

Greene, R.F. and R.W. O'Donnell, 1966: Phys. Rev. **147**, 599. Sect. 3.2

Häupl, K. and P. Wißmann, 1984: Z. Naturforscher **39a**, 481. Sect. 4.2.2

Heimlich, C., 1983: Diplom thesis, University of Kassel F.R.G. Sect. 5.4

Heinrichs, J., 1973: Phys. Rev. B **8**, 1346. Sect. 3.2

Henderson, B., 1972: *Defects in Crystalline Solids*, (Crane, Russak and Company, Inc., New York) p.139. Sect. 5.1

Hinch, B.J., C. Koziol, J.P. Toennies and G. Zhang, 1989: Europhys. Lett. **10** 341. Sect. 5.4

Hoffmann, H. and G. Fischer, 1976: Thin Solid Films **36**, 25. Sect. 5.4

Hoffmann, H. and P. Kücher, 1987: Thin Solid Films **146**, 155. Sect. 5.4

Holzapfel, C., 1988: Diplom thesis, University of Düsseldorf F.R.G. Sect. 5.3

Holzapfel, C., F. Stubenrauch, D. Schumacher and A. Otto, 1990: Thin Solid Films **188**, 7. Sect. 5.1

Holzapfel, C., W. Ackemann, D. Schumacher and A. Otto, 1990: Surface Sci. **227**, 123. Sects. 5.1, 5.3

Hölzl, J. and F.K. Schulte, 1979: *Work Funktion of Metals in Solid Surface Physics* in: Springer Tracts in Modern Physics, **85**, 1 (Springer, Berlin, Heidelberg, New York). Sect. 5.2

Horne, J.M. and D.R Miller, 1977: Surface Sci. **66**, 365. Sect. 3.2

Hove, M.A. van and S.Y. Tong, 1979: *Surface Crystallography by LEED* (Springer, Berlin, Heidelberg, New York). Sect. 3.2

Jalochowski, M. and E. Bauer, 1988: Phys. Rev. B **38**, 5272. Sect. 5.4

Jonker, B.T., N.C. Bartelt and R.L. Park, 1983: Surface Sci. **127**, 183. Sect. 5.4

Jonker, B.T. and R.L. Park, 1984: Surface Sci. **146**, 93; Solid State Communications **51**, 871. Sect. 5.4

Juretschke, M.J., 1965: J. Appl. Phys. **37**, 435. Sect. 3.2

Kannen, G., B. Reihl and A. Otto (to be published). Sect. 5.3

Karaus, A., 1984: Diplom thesis, University of Düsseldorf F.R.G. Sect. 3.1

Keppner, W., R. Wesche, T. Kas, J. Voigt and G. Schatz, 1986: Thin Solid Films **143**, 201. Sect. 5.4

Klug, P.H. and L.E. Alexander, 1954: *X-Ray Diffraction Procedures* (J. Wiley and Sons, New York). Sect. 4.2.2

Koch, J.F. and J.D. Jensen, 1969: Phys. Rev. **184**, 643. Sect. 3.1

Kossel, W., 1927: Nachr. Ges. Wiss. Göttingen, Math. Phys. Kl. **135**. Sects. 4.2.1, 5.2

Koziol, C., G. Lilienkamp and E. Bauer, 1987: Appl. Phys. Lett. **51**, 901. Sect. 5.4

Landolt-Börnstein, 1959: *Zahlenwerte und Funktionen 6.Teil, Elektrische Eigenschaften I*, Ed.: K.H. Hellwege and A.M. Hellwege, (Springer Verlag, Berlin, Göttingen, Heidelberg). Sect. 2.3

Lang, N.D., 1973: Solid State Physics **38**, 225, Ed.: F. Seitz, D. Turnbull and H. Ehrenreich, (Academic Press, New York). Sect. 3.1

Lessie, D., 1979: Phys. Rev. B **20**, 2491. Sect. 3.2

Lessie, D. and E.R. Crosson, 1986: J. Appl. Phys. **59**, 504. Sect. 3.2

Leung, K.M., 1984: Phys. Rev. B **30**, 647. Sect. 3.2

Linde, J.O., 1931: Ann. d. Physik **10**, 52. Sect. 5.1

Linde, J.O., 1932: Ann. d. Physik **14**, 353. Sect. 5.1

Lucas, M.S.P., 1964: Appl. Phys. Lett. **4**, 73. Sects. 2.4, 5.1

Lucas, M.S.P., 1965: J. Appl. Phys. **36**, 1632. Sect. 3.2

Lucas, M.S.P., 1968: Thin Solid Films **2**, 337. Sects. 2.4, 3.3

Marliére, C., 1986: Thin Solid Films **136**, 181. Sect. 5.4

Mayadas, A.F., M. Shatzkes and J.F. Janak, 1969: Appl. Phys. Lett. **14**, 345. Sect. 2.3

Mayadas, A.F. and M. Shatzkes, 1970: Phys. Rev. B **1**, 1382. Sects. 2.3, 2.4

McCarroll, B. and G. Ehrlich, 1963: J. Chem. Phys. **38**, 523. Sect. 5.4

Meinel, K., M. Klaua and H. Bethge, 1988: J. of Crystal Growth **89**, 447. Sect. 5.4

Miguel, J.J. de, A. Cebollada, J.M. Gallego, J. Ferron and S. Ferrer, 1988: J. of Chystal Growth **88**, 442. Sect. 5.4

Mitchinson, J.C. and R.D. Pringle, 1971: Thin Solid Films **7**, 427. Sect. 3.3

More, R.M. and D. Lessie, 1973: Phys. Rev. B **8**, 2527. Sect. 3.2

Munitz, A. and Y. Komem, 1980: Thin Solid Films **71**, 177. Sect. 5.2

Namba, Y., 1970: Jap. J. Appl. Phys. **9**, 1326. Sect. 3.2

Nee, T., J.F. Koch and E. Prange, 1968: Phys. Rev. **174**, 758. Sect. 3.1

Norbury, A.L., 1921: Trans. Far. Soc. **16**, 570. Sect. 5.1

Pariset, C. and J.P. Chauvineau, 1975: Surface Sci. **47**, 155. Sect. 2.4

Pariset, C., 1976: Thesis, Paris, CNRS AO 9714 (F). Sects. 2.4, 5.4

Pariset, C. and J.P. Chauvineau, 1976: Surface Sci. **57**, 363. Sect. 2.4

Pariset, C. and J.P. Chauvineau, 1978: Surface Sci. **78**, 478. Sects. 2.4, 5.1, 5.4

Pariset, C., M. Gasgnier and M. Galtier, 1975: Thin Solid Films **29**, 325. Sect. 2.4

Parrish, W., 1962: *Advances in X-Ray Diffractometry and X-Ray Spectrography*, (Centrex Publishing Company, Eindhoven). Sect. 4.2

Pendry, J.B., 1974: *Low Energy Electron Diffraction* (Academic Press, London). Sect. 3.2

Perdereau, J., J.P. Biberian and G.E. Rhead, 1974: J. Phys. F **4**, 798. Sect. 5.4

Persson, B.N.J., 1991: Phys. Rev. B **44**, 3277. Sect. 3.2

Persson, B.N.J., D. Schumacher and A. Otto, 1991: Chem. Phys. Lett. **178**, 204. Sects. 3.2, 5.1

Persson, B.N.J. and N.D. Lang, 1982: Phys. Rev. B **26**, 5409. Sect. 3.1

Pichard, C.R., C.R. Tellier and A.J. Tosser, 1979: Thin Solid Films **62**, 189. Sect. 2.3

Pichard, C.R., C.R. Tellier and A.J. Tosser, 1980: Phys. Stat. Sol. (b) **99**, 353. Sect. 2.3

Poelsema, B. and G. Comsa, 1989: *Scattering of Thermal Energy Atoms* in Springer Tracs in Modern Physics, **115**, 1. Sect. 3.2

Polomski-Keip, H.J., 1986: Diplom thesis, University of Düsseldorf F.R.G. Sect. 3.1

Potter, R.J. and D. Dexter, 1957: Phys. Rev. **108**, 677. Sect. 5.1

Powder Diffraction File, 1974: Inorganic Sets, 1-5, 4-783, Joint Commitee on P.D. Standards, 1601 Park Lane, Swathmore, Pennsylvania 19081, USA. Sect. 4.2.2

Prange, R.E. and T.W. Nee, 1968: Phys. Rev. **168**, 779. Sect. 3.1

Roth, C., 1989: Diplom thesis, University of Düsseldorf F.R.G. Sect. 5.2

Roth, C. and D. Stark, 1990: Verh. DPG, DS **11.2**, 655. Sect. 5.2

Roth, C., D. Schumacher and D. Stark (to be published). Sects. 5.1, 5.2

Sambles, J.R., 1983: Thin Solid Films **106**, 321. Sect. 2.4

Sambles, J.R., 1987: Thin Solid Films **151**, 159. Sect. 3.2

Sambles, J.R. and K.C. Elsom, 1982: J. Phys. D **15**, 1459. Sect. 3.2

Sandomirskii, V.B., 1967: Sov. Phys. JETP **25**, 101. Sects. 2.2, 5.4

Scherrer, P., 1918: Gött. Nachr. **2**, 98. Sect. 4.2

Schlemminger, W., 1982: Diplom thesis, University of Düsseldorf F.R.G. Sect. 4.2.2

Schlemminger, W., 1989: Thesis, University of Düsseldorf F.R.G. Sects. 4.1, 4.2.2

Schlemminger, W. and D. Stark, 1986: Thin Solid Films **127**, 49. Sect. 4.1

Schulte, F.K., 1976: Surface Sci. **55**, 427. Sects. 2.2, 5.2

Schulte, F.K., 1977: Phys. Stat Sol. (b) **79**, 149. Sect. 5.4

Schumacher, D., 1983: Thesis, University of Düsseldorf F.R.G. Sect. 5.4

Schumacher, D. and D. Stark, 1982: Surface Sci. **123**, 384. Sects. 5.1, 5.4

Schumacher, D. and D. Stark, 1983: Verh. DPG, DS **10**, 550. Sect. 5.3

Schumacher, D. and D. Stark, 1984: Thin Solid Films **120**, 15. Sect. 5.4

Schumacher, D. and D. Stark, 1986: Thin Solid Films **139**, 33. Sect. 4.1

Schrammen, P. and J. Hölzl, 1983: Surface Sci. **130**, 203. Sect. 5.2

Soffer, S.B., 1967: J. Appl. Phys. **38**, 1710. Sect. 3.2

Sondheimer, E.H., 1952: Advances in Physics **1**, 1. Sects. 2.4, 3.2, 5.3

Spiller, G. and A. Segmüller, 1989: Ann. New. Acad. Sci., **342**, 188. Sect. 5.4

Stark, D., 1987: Surface Sci. **189/190**, 1111. Sect. 5.2

Stranski, I.N., 1928: Z. Phys. Chem. **136**, 259. Sect. 5.2

Stubenrauch, F., 1987: Diplom thesis, University of Düsseldorf F.R.G. Sect. 5.3

Suhrmann, R., 1957: J. Chem. Phys. **54**, 15. Sect. 5.1

Tanaka, H. and L. Boesten, 1985: XIV Int. Conf. on Electron-Atom Collision, Palo Alto, p.237. Sect. 5.3

Tellier, C.R., C.R. Pichard and A.J. Tosser, 1979: Thin Solid Films **61**, 349. Sect. 2.3

Tellier, C.R. and A.J. Tosser, 1982: *Size Effects in Thin Films*, Thin Films Science and Technology 2, (Elsevier Scientific Publishing Company, Amsterdam, Oxford, New York). Sect. 2.3

Thomson, J.J., 1901: Proc. Cambridge Phil. Soc. **11**, 120. Sect. 2.4

Tonscheidt, A., 1990: Diplom thesis, University of Düsseldorf F.R.G. Sects. 5.1, 5.2

Trivedi, N. and N.W. Ashcroft, 1988: Phys. Rev. **38**, 12298. Sects. 2.2, 5.4

Vand, V., 1942: Proc. Phys. Soc. **55**, 222. Sect. 4.1

Voort, E. van der and P. Guyot, 1971: Phys. Stat. Sol. (b) **47**, 465. Sect. 2.3

Vries, J.W.C. de, 1987: Thin Solid Films **150**, 201. Sect. 2.3

Washburn, S., H. Schmid, D.P. Kern and R.A. Wepp, 1987: Phys. Rev. Lett. **59**, 1791. Sect. 2.2

Warkusz, F., 1988: Thin Solid Films **161**, 1. Sect. 2.3

Watanabe, M., 1973: Surface Sci. **34**, 759. Sect. 3.2

Watanabe, M. and A. Hiratuka, 1979: Jap. J. Appl. Phys. **18**, 31. Sect. 3.2

Wedler, G., 1987: *Adsorption and Reaction on Thin Metal Films* in Thin Metal Films and Gas Chemisorption, Ed.: P. Wißmann, Studies in Surface Science Catalysis, Vol. 32 (Elsevier, Amsterdam, Oxford, New York, Tokyo). Sects. 2.4, 5.1, 5.3

Wepp, R.A., S. Washburn, C.P. Umbach and R.B. Laibowitz, 1985: Phys. Rev. Lett. **54**, 2696. Sect. 2.2

Wißmann, P., 1975: *The Electrical Resistivity of Pure and Gas Covered Metal Films* in Springer Tracts in Modern Physics, **77**, 1. Sects. 2.4, 3.2, 4.2.4, 5.1

Wittmann, E., 1984: Thesis, University of Erlangen-Nürnberg F.R.G. Sect. 5.1

Zhang G., 1991: private communication. Sect. 5.4

Ziman, J.M., 1960: *Electrons and Phonons* (Clarendon Press, Oxford). Sect. 3.2

Zinsmeister, G., 1966: Vakuum **16**, 529. Sect. 5.4

Subject Index

Springer Tracts in Modern Physics

* denotes a volume which contains a Classified Index starting from Volume 36